INTO THE LOST WORLD

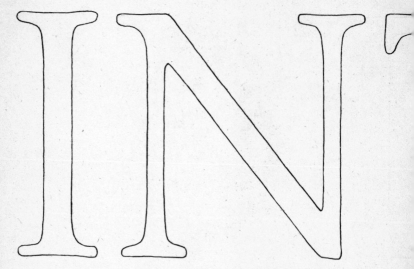

Also by David Nott: ANGELS FOUR

THE
LOST
WORLD

DAVID NOTT

PRENTICE-HALL, INC., *Englewood Cliffs, N.J.*

Photo credits: Charles and James Brewer-Carías, Hernan Richardo, and José Luis Blasco

Design by Linda Huber

Into the Lost World by David Nott

Printed in the United States of America

Prentice-Hall International, Inc., London
Prentice-Hall of Australia, Pty. Ltd., Sydney
Prentice-Hall of Canada, Ltd., Toronto
Prentice-Hall of India Private Ltd., New Delhi
Prentice-Hall of Japan, Inc., Tokyo

10 9 8 7 6 5 4 3 2 1

Library of Congress Cataloging in Publication Data

Nott, David.
    Into the lost world.

    1. Expedición de los Tepuís Venezolanos, 1974. 2. Sarisariñama, Venezuela. 3. Natural history—Venezuela—Sarisariñama. 4. Geology—Venezuela—Sarisariñama. I. Title.
QH11.N6   508.87'6   74-26691
ISBN 0-13-477190-7

Acknowledgements

CHARLES BREWER-CARÍAS DESERVES THE GRATITUDE of all of us for dreaming up, planning and pushing this big expedition project through to a successful end. It has been suggested by members of the expedition that the major hole in the Sarisariñama plateau be named Sima Brewer (sima-pit), and the smaller one Sima Gibson, in recognition of the pilot Harry Gibson, who first saw them from the air in 1964. This will of course depend on the laws binding the Cartographic Division of the Ministry of Public Works.

My personal thanks go to Washington journalist Jessica Watson, wherever she may be, for her writer's eye, for exorcising the more exaggerated anglicisms from the manuscript, and for constantly needling me to get on with the job.

To
ANNABELLA GWYNEDD NOTT

1

THE BRIGHT RED AND WHITE HELICOPTER LIFTED OFF
from Camp One on top of the plateau and clattered
north, skimming the jungle which covered its surface.
Looking down through the open door at the trees,
bending and flailing beneath us under the downdraft of
the rotor, twigs and leaves flying, we got the impression of
a boiling green caldron, like heavy seas crashing and
swirling in a rocky inlet. Three of us were about to drop
into that caldron from about 60 feet up.

Then, abruptly, we shot over the edge of the great hole
sunk in the plateau. The hole that we were supposed to
climb down and explore . . . The Pit. In that instant the
drop changed from 60 to nearly 1,000 feet. We got a
flashing glimpse down the immense, vertical rock walls of
the hole to the mysterious jungle at the bottom and then
we were across its 400-yard diameter and banking
sharply to take another run in.

Two more passes and we got the pilot lined up on a red-flowering tree growing on the surface of the plateau near the point on the edge of the hole where we reckoned we could begin our climb down into it. We circled once again and the helicopter slowed and hovered.

Charles, the first to go, went out the door and swarmed down the rope ladder. Watching from above, I saw that the ladder ended way above the jungle floor. Worse, it didn't even reach the top of the trees. *For God's sake*, I seethed, *we haven't even started on the lig climb down into the hole and we get this little surprise thrown at us!*

I watched Charles swinging at the end of the ladder, one foot on the bottom rung, the other waving in the empty air beneath it, looking up and around him in consternation. There was of course, no communication; the infernal din of the chopper saw to that. But Charles came through. We had not been working ourselves up for three months for this expedition for nothing. He gave one last enquiring glance upwards and simply let go. He dropped about 15 feet, crashed through the frail top-twigs of a tree until the firmer branches stopped him, hopelessly entangled, his blue helmet askew. *That's what happens when you don't rehearse an unknown maneuver*, I told myself for the umpteenth time in an expedition career full of such mishaps. But we hadn't rehearsed and it was too late now. Jimmy went out next and I wagged a finger at him, trying to convey, as if he didn't know, that if he missed a tree in his drop he would hit the jungle floor with a probably fatal bang. I didn't watch him go. I had other jobs to do. I uncoiled an old 130-foot, 14 mm-thick climbing rope I had brought along for that purpose, clipped it to a safety ring on the floor of the helicopter, snaplinked our five duffel bags of kit to it and slid them down. Then I decided to go down the rope myself. To hell with that skimpy ladder. As I ducked out the door I noticed that Jimmy had landed in a tree all right, but it wasn't the same one. He was a good 30 feet away from Charles. *Now what's that all about*, I said to myself as I dropped out. Helicopters don't hover in one place. They move around a bit. I'd dropped my rope carefully between trees to the floor of the jungle. But that's not where I landed. I hit a tree too. A third one. So there we were, the three great expeditionaries, each ludicrously entwined in the top branches of his own tree, breaking off twigs, with their black, slimy bark, to free our limbs, and dropping, a yard or so at a time, to the top of the main trunk. Once we got firmer timber in our hands we shinned down to the ground and watched as the helicopter clattered away.

Nothing leaves a silence behind it like one of those infernally noisy machines, and an isolation too, in our case, because it would be six days before it came back for us, if, that is, we got safely down into The Pit, and once down there, were able to get out again.

We brushed ourselves off, extracting leaves and twigs and wood splinters from our ears and from down our necks, and surveyed each other. We were wearing tough canvas overalls with long sleeves and reinforcing patches on the behind, the knees and elbows. Charles's was orange, Jimmy's white and mine blue so that in the aerial filming of our descent it could be seen who was where, and doing what. The overalls were already stained and streaked with the green and black goo off the trees. Over them we had on our climbing harness and looked like a bunch of fancy parachutists, which we all were in fact, who had fallen off a truck. No one had said a word and I broke the silence with one of my endless and, to others, maddening timechecks. "It's 9:00 A.M." I said. The others didn't reply. They knew what I meant was we were two hours late and we had an unknown, nearly 800-foot wall to get down and a bivouac to find at the bottom, before nightfall. We hunted around for our packs, and then started to drag them through the tangle towards The Pit.

If you haven't been in jungle or tropical forest it's difficult to envision. In the movies, Tarzan swings from giant tree to tree to land in a nice, grassy clearing. There are no clearings. And you couldn't swing more than a few feet anyway because things are so thick the sunlight itself has a hard time getting through. But what matters to you, the explorer, is the undergrowth, which blocks your view of anything more than three feet away, let alone your passage; and even more, the surface underfoot, or more accurately, the lack of it. You can crawl scores of yards from one rotting, fallen tree to the next, careful not to put your full weight onto any one of them, and never touch the thick squelchy tangle of roots and earth and humus which lies beneath. Not unless you fall through, of course, in which case you climb out wondering yet again how come you didn't snap a bone down there between the branches. Now you have to add to this the weight and the sheer cussedness of your packs. It's my practice to pick them up—70 to 100 pounds each—and throw them ahead so they break through two or three feet of undergrowth, step up to them, pick them up and throw them ahead again. If you are using a machete you cut three feet and then have to reach back awkwardly for the packs and drag them forward. I prefer the weight lifting.

Dude expeditions are another matter. They have a

machete man in front with his hands free, the dudes next with their cameras, and the laden porters behind. But we were three men humping climbing gear, caving gear, bivouac kit, and supplies to last six days. And the machete was in one of the packs because you don't climb down a major wall with one of those hanging from your harness. We considered this rough going. It was nothing compared to what awaited us at the bottom of the hole.

We fell into our own sequence. It was my job, as the expedition's climber, to equip the route down the wall so I went ahead with the pack containing five 9-millimeter thick, 195-foot long ropes, nearly 1,000 feet in all, and all the rock climber's ironmongery—pitons, which are steel blades you drive into cracks for security, hammers, slings, snaplinks and so on—dangling from my harness (and catching on every snag and twig in the forest). The others followed with two heavy packs each.

We advanced by degrees until brighter light in the foliage ahead of us told us we were nearing the edge of The Pit. Almost immediately we came to a 40-foot wall of rock dropping vertically down in front of us. It had nothing to do with the main wall. It was just one of those little traps and obstacles that The Pit had in store for us. I selected a reasonable tree, tied on our extra rope, the one I had slid down from the helicopter, and dropped it down the wall. I clipped two turns of it into a snaplink on my harness and rappelled down it. In a rappel, you slide down the rope controlling the speed of your descent by the friction of the rope through the snaplink and against your thigh. They are even doing it in movie thrillers nowadays and think they have a smart new gimmick.

The others dropped the packs down the rope and

followed. I moved on down through steeply sloping jungle until, brushing away the foliage in front of me I got my first look across The Pit to the other side, level with me and about 400 yards away. I was lying flat, crawling forward until suddenly I reached the edge and peered straight down 750 feet of vertical rock wall to a luridly green bowl of jungle at the bottom. The great amphitheater of walls, huge sections of them overhanging, was unique in my experience. I'd never seen a complete circle of rock walls and I thought of the Colosseum in Rome, tiny by comparison, where so many had entered its arena and never left alive. Here we were struggling to get into our own arena, and with no slavemaster's sword at our backs to force us on. This tomfool idea was discouraging enough but it rapidly evaporated when a new factor in my immediate life became apparent. Shifting slightly I found that the world was bouncing gently like a trampoline. I shifted again and up and down I went, faintly but unmistakably, in diminishing bounces until the world came to rest again.

And then I understood. I wasn't on terra firma at all. I was lying on a cornice of rotting humus, earth and roots, two yards deep and jutting out eight feet over the drop. I was on a crumbling ledge sticking out from the seventy-fifth floor of the Empire State building. Very slowly I crept backwards, a foot at a time. With each movement the bouncing began and I waited without breathing for it to subside before moving again. I supposed the cornice would break away with a ripping, tearing sound of roots and I listened hard for the first of them to give way. I heard little cracks and snappings and decided to accelerate. As in the lives of most of us, I've been told now and then that I'm a bit of a snake. Well, I'm the first snake that's ever gone into reverse. My backward crawl must have topped seven miles an hour until I reached the first tree, jumped up and crashed through the vegetation to meet the inquiring stares of Charles and Jimmy. "Wrong place," I muttered.

Looking down from the cornice, I had noticed, before the bouncing got through to me, that straight below me the wall was blank for several hundred feet. Further to the right, however, there was a grassy ledge about 150 feet down. This could be a landing place for our first rappel. Dragging the rope-bag and pulling the end of the extra rope behind me I moved to the right until I judged I must be at a point above the ledge. Here I fished out the first of the new, red ropes and joined it behind a tree to the old rope with a bowline. I was moving too fast for caution, perhaps because I knew that in a few moments I would be bouncing around on a cornice similar to the first one and, moreover, would this time have to go over its edge. We had already lost another hour, so I stepped quickly onto the cornice, disdaining to notice the bouncing now that I had a rope in my hand, perched myself on its edge and, very carefully separating the coils of the rope into a bunch in each hand, threw one over the drop. The rope curved out in a beautiful arc and soon the coils were running swiftly out of my other hand. Two hundred feet out in one clean throw. Pleased, I clipped the rope twice through the snaplink, took a firm grip of it across my thigh and stepped backwards into space. I was expecting the first lurching drop because I was well used to the elasticity of nylon rope. But now there were two rope lengths joined above me, and the first ran in a diagonal from the first rappel point over on the right through the undergrowth. I dropped about 12 feet, swinging underneath the cornice and watching it bounce over my head in tune to my own bouncing movement until my weight drew out the elasticity of the ropes.

My first thought ran like a headline in my brain: "A SHEETBEND IS RECOMMENDED FOR JOINING TWO ROPES OF UNEQUAL THICKNESS." The red rope was 9mm thick and the old rope 14, and I'd used a bowline. As I pondered this lesson from the book of knots I twirled slowly on the rope until I was facing away from the rock and into the hole. From my position, suspended in space, the view was magnificent. The whole great Pit opened around and beneath me. I felt like an eagle, or at any rate, a Yo-Yo, and forgetting knots and splices I let out the first human yell that place had ever heard in all history. Then I slid steadily down in an exhilarating free drop to the ledge, out of contact with the rock, twirling continually on the rope, and nearing the end of it, I swung across to a minute tree growing in a crevice and got myself balanced on a thin line of grass above it. First stretch completed. Six hundred feet to go.

I shouted up for the packs to be lowered and waited gleefully for the other two to take the plunge over the Bouncing Cornice. As the packs reached me I bent down carefully and clipped them onto the little tree. With two there I paused to light a cigarette. At this moment one of the packs broke free from the rope above and whooshed past my head to crash down the wall below. For some reason or other I simply shrugged to myself and struck another match. The film team with telephoto lenses who had been helicoptered to a point across the hole from us, later told me they'd noticed I'd needed two matches to light that cigarette and I promptly launched the legend that the first had been blown out by the passing bag. But a sudden thought stripped me of such commendable nonchalance. The ropes. Suppose what had fallen was the rope-bag? That would be the end of the expedition. An irritated No floated down in answer to my bellowed questions. But I was the one with the right to be irritated. That pack had all but killed me. So I followed up with some deft reminders about how the Bouncing Cornice would straighten out the ideas of the idle scoundrel who had tied the bag on and almost knocked me off the rock.

Two more packs came down and then a squawk of alarm. I looked up to see Jimmy dropping over the overhang and swinging giddily under it. He rappelled slowly down to join me, breathless, on the perch.

"What do we tie onto?" he asked, appalled at our position.

"Nothing, Jimmy. There's no crack for a piton."

I pulled another rope out of the bag, tied it onto the one we had just come down, leered at him, and slid off our little grass tuft. "When I shout to you that I've reached another perch, you tell Charles to come down to you. Then lower the packs to me and follow them down," I said. I shouldn't have been so perky; this wall was full of tricks.

About 30 feet lower down, I spotted the fallen bag over to the right stuck in a tree that was growing out of a crevice. There was no question about leaving it there, of course. I had to go and get it. It was about 15 feet away from me and at the same level. I studied the rock between myself and it, looking for a line of holds across which I could traverse to reach it. But the wall was uncompromisingly vertical and smooth. I could climb back up the rope and try to traverse higher up until I was directly above the tree and then rappel down to it, but this would alter the route of descent and the whole process would have to be reversed to get us on the right line again.

The trick to solve this situation was a pendulum. This, at first sight, is a maneuver as lunatic as a motorbike roaring 'round the Wall of Death. The position is much the same too. The wall is vertical and the rider and his machine horizontal; the rock is vertical and the climber horizontal. The rider knows, of course, that centrifugal force will keep him on his wall; the climber is held up by his rope. I turned until my right shoulder was pointing down the drop and ran a few paces across the face. I swung back again past my starting point and then, with the force of the swing with me, sprinted across, let go the rope with one hand and grabbed the tree. It was a flimsy plant, perhaps two inches thick at the root. As I reached up and outwards for the bag, the already battered branches broke off and I swung back across the rock out of control, one hand on the rope and the other gripping the bag. A real circus trick over a 600-foot drop and without a net.

It's hard enough to hold your own weight with one hand on a wet 9mm rope, which is not much more than a quarter-inch thick, and the weight of the bag was too much for me. *It's me or you*, I told it as we reached the end of the swing, and I let it go, willing it to stop on the next ledge about 40 feet down and over on the left. But I had no time to watch its fall because I had another problem. My rope, which should have hung vertically below me from the snaplink, was jammed in the only fork left in the remains of that same tree, and the same 15 feet away on the right. My hands and arms were about to give out by now (physical training buffs might like to know this was despite ten fingertip pull-ups, five times a day for three months) and the 35-minute struggle which followed before I extricated myself became quite serious. The only organs really working were my lungs (gasps) and my tongue (curses). But those interested in the spiritual solidarity and brotherhood that grows between men in danger, will be pleased that the whole scene was enlivened by catcalls from above and ribald queries about what had happened to the macho who shows off smoking cigarettes on holdless perches.

I finally reached the next ledge, noted without elation that the pack had indeed stopped there, and promptly sat down on it. *The other two can sort themselves out*, I thought, reaching for a cigarette. I should have savored it more appreciatively, for it was to be my last smoke on the wall.

It was some time before the others joined me on the ledge and we held a rapid council of war. Time, as always, was the enemy. We were only a third of the way down the route and the setting sun already lit only the eastern arc of the hole. The rest was in shadow and the bottom in gloom. But below us our line eased off the vertical, and we could see that the vegetation was thicker, eventually becoming real jungle again but at so steep an angle it seemed unbelievable that trees could find purchase there. We rappelled down a few more walls and wet, mossy chimneys until we found the jungle made our climbing methods impossible. I claim I can throw a coiled rope in a curve that will take it around a corner. But there was no way to throw it down through the tangle that was closing in on us. We were on very steep, loose humus giving way without warning to vertical drops and gullies. There was only one way to deal with it. I called it Operation Rump. We would sit down, rope in one hand, two bags joined with a loop in the other, and toboggan. I would go first with the rope-bag and lead out the safety rope, stopping at each vertical drop to fix up a proper descent. Our progress was hilarious. Three men and five bags careening down on their backsides in an avalanche of dirt, shooting over bumps, vanishing into prickly hollows, swerving round trees and collecting more cuts and bruises in half an hour than a Welsh rugby forward gets in a month. Every so often I would scream a warning and we'd all brake hard on the rope and wrestle our 130-pound loads to a stop, right on the edge of one of The Pit's myriad traps. But there was another trap closing in on us. Darkness.

As each hour of descent went by, the circle of sky above us, framed by the rim of the hole, grew smaller; the air more still and dank. In the deepening gloom we were shocked by the rumble of a rock-fall bouncing round the walls in a minutes-long echo that seemed to spiral upwards to flutter into silence far above. Beneath us there was nothing now to show the route we should follow. We knew there were big walls at the bottom which we should avoid by trending rightward. But we knew too that if we went too far that way we'd end up traversing out across the wall over nothing again. We were in fact beginning to get the picture. The Pit was a malevolent place and we, the first men in it, would not be forgiven any mistakes.

On and on down we went. Rope after rope. Probing

false lines and climbing wearily back up again dragging those confounded packs every inch of the way. We continued rightward in a descending traverse, feeling, rather than seeing through the foliage, the presence of the wall. By now the last rays of sunlight which had lit the top rim of the eastern wall had gone and we had reached the sloping curve of the great bowl of jungle at the bottom. Instinctively I turned away from descending further into that unknown basin in the dark and cut up rightward again.

"Where are you going?" Jimmy said.

"Over to the bottom of the wall."

"Why?"

"Because it's the right place to be."

"Like hell it is. What about that rock-fall? You want to get us killed?"

I almost cracked him one with my piton hammer. We were getting short-tempered, as tired men do in a doubtful situation. I'd seen a huge overhang about 150 feet up from the bottom of the wall. If we got under that we'd be safe from falling rock. It would also be dry. In the spot where we were, there wasn't a level square yard of ground and it was covered with festering undergrowth too. Also, if it rained, we'd be sluiced out of our bivouac in minutes. I pushed on across the slope and could soon see the pale gray glimmer of the rock ahead. In the last flicker of light we reached it, and a dry level stretch of ground in the six-foot space between the trees and the wall.

The descent had taken nine hours, without food and above all without water. But we made no move to dig out the water canteens from the packs or the food either. We had something else on our minds. Something shelved and purposely forgotten in the stress of the descent. We sprawled in the darkness for some time without speaking. And then:

"I know what you're thinking," I said. "You don't like the idea of going up that last 150 feet of rope at the top of the wall. Under the cornice."

No, they didn't. This ascent would have to be made with jumars, metal devices which clip onto the rope and which will slide up it but lock onto it when you apply downward pressure. The climber stands in loops of nylon cord or tape attached to the jumars and moves up the rope a step at a time. It can be exhausting and needs a lot of nerve over a long drop. Charles said: "What's more, our weight, coming fully on the rope when we rappelled down it, buried it deep in the damned cornice. We can't jumar over it."

So you already know, I thought to myself. I spoke up. "Right. But that's not the only problem. That cornice might break away when one of us is on the rope beneath it."

There was another silence. And then Jimmy chimed in. "You mean we can't get out?"

"That's right."

"You hear that, Charles?" he said incredulously. "We're trapped."

2

THE DREAM OF DISCOVERING A LOST WORLD DIES HARD. No matter what science says, men still maintain a flicker of hope of finding it. The dream began with the famed tale by Sir Arthur Conan Doyle in which four English explorers climb to the top of a remote plateau in South America to find that prehistoric life, with its dinosaurs and pterodactyls, has survived up there when it had long become extinct on the rest of the planet. Their adventures include clashes with cavemen and snatching fist-sized diamonds from under the great jaws of flying reptiles.

The models for the plateau in this story are to be found in the largely unexplored regions of south and southeast Venezuela. The Auyán-Tepuí and Roraima are two which were explored but which failed to deliver up any of the legendary monsters. In one area down there, about 500 miles southeast of Caracas, the capital, and bound by

15

latitudes 4 degrees and 5 degrees north, and longitudes 63 and 65 degrees west, is the source of the Caura, Erebato and Ventuari Rivers, the first being the largest tributary of the Orinoco. Near the southern reaches of the Caura, barely 20 miles from the Brazilian frontier, a similar plateau, 5,000 feet high, rises sheerly from the jungle. It is 25 miles wide, about 15 miles from north to south and its surface is covered with thick forest. In the lowland jungle to its south near the Canaracuni River live a sprinkling of primitive Sanema Indians. Another tribe, the Makiritare, have also seen the plateau and given it the magic name Sarisariñama. It was unknown, untouched, unexplored. Could this be the Lost World?

Each of these plateaus is an island, isolated from its neighbors by mile upon mile of unbroken jungle, and even from the lowland jungle beneath it by its own great rock walls, some of them several thousands feet high. The summits are comparatively cold, covered for weeks at a time with dense clouds, glimmering with tropical lightning, and offer varied terrain: barren rock split by deep crevasses, swamps, wildly sculptured rock towers and pinnacles that seem to move in the shifting mist, scrubland, dwarf forest and areas of moderately high trees reaching about 30 feet. But what draws the scientist is that the conditions prevailing today—of solar radiation, altitude, humidity, soil nutrients and so on—have apparently changed very little since the first forms of life appeared there millions of years ago. This has allowed the survival of species on the plateaus which are extinct everywhere else. They are called ancient relict flora and fauna. Moreover, some original species, isolated on these summits, underwent a series of mutations dictated by the microclimates and other conditions peculiar to each plateau and have evolved into what are called endemic species. In other words, they exist only on the summit on which they are found and nowhere else on earth.

The plateaus are of sandstone and they belong geologically to the Guiana Shield, which covers this area of the South American continent and is one of the oldest rock formations in the world, dating back perhaps as much as 3,800,000,000 years. Life was thriving on top of their summits millions of years ago when the oceans rolled in and submerged the great plains of the region. While landlife was extinguished over vast territories, it survived on these highlands. It is the same now as it was before the infinitely slow encroachment of the oceans began; it continued undisturbed during the eons of time the inland seas remained, and through the ages of their slow retreat. The plateaus are islands in time.

The scientists see them all as treasure islands too. But they were particularly attracted to Sarisariñama because aerial photos showed that the forest which covers its summit was thicker than on other plateaus, and richer than would be expected from such acid and rocky ground. But this was only the beginning of the mystery which drew us to it. There was something else.

In 1964 veteran Venezuelan bush-pilot Harry Gibson was flying, by chance, across the plateau when, astonished, he stood his plane on its wingtip in a tight turn to check whether he had really seen what he thought he'd seen or whether it had been a trick of his imagination. It was The Pit. And about a mile to the south of it another, smaller hole, sunk inexplicably in the flat jungled surface of the plateau, its pale gray and yellow walls startling in this endless world of green. Harry banked over the great hole again to get a look straight down to the bottom.

There was jungle there too! A jungle of its own. More luxuriant, taller, and a deeper green than the jungle on the surface.

When the news reached the scientists these normally cautious and guarded men were agog. Here was a TIME CAPSULE within an ISLAND OF TIME! What extraordinary discoveries were hidden in the strange jungle at the bottom of The Pit? The laymen were quick to reflect the excitement in their own terms: "If there's a pterodactyl left on earth, it's got to be down that hole. If the Lost World exists, it has to be there," they announced.

Thus all the ingredients for a big spectacular were there. But the location was so remote and the approach so difficult that what was missing for many years was the man to produce it. The man, finally, was Charles Brewer-Carías, thirty-five, a Venezuelan and very proud of it, but as far from looking like one as his name would suggest. He is of medium height, blond, with a droopy gold mustache. He has seven international swimming medals and has led five large-scale jungle expeditions, one of which he produced in a book on the dental anthropology of the Soto Indians, thus keeping his profession as a dentist in tune with his vocation as an explorer. He has organized a similar number of forays into the Venezuelan Andes, is a master parachutist and once led a jump of ten men into the headwater region of the Erebato River on a survival course of nine days. Equipment: one machete and two matches.

Above all Charles is a determined man. We have been on three unlikely adventures together, and I have come to trust his ability to produce those far-out items of the

explorer's dreams—air reconnaissance, heavy air transport, helicopters, airlifts of fuel to remote jungle airstrips,
and the like. He trusts me as a veteran climber to get up,
down or across any mountain or precipice that may be in
the way. We make a good team. He does all the work,
which for a major expedition is formidable, and I get
airlifted or dropped in front of a problem, size it up and
take a run at it. It should be the other way 'round
because I am ten years his senior. But there it is. That's
the way we prefer it.

He told me about The Pit some years ago but it didn't
sink in. Then I lost long months with fractures from a
parachute accident, and we later went our separate ways
with different expeditions. I had another bout with
broken bones from my newfound sport of motocrossing
and was still limping a little when he inveigled me to his
house, pointed to a three-dimensional spectroscope with
a color slide beneath it and suggested I take a look. It was
The Pit. In 3-D, full-color Panavision, or something. All
that was needed was a Wagner tape at full blast. I was
heartstruck. To think that frightening geological freak
had been there all this time waiting for the likes of us,
and we two had been foraging around somewhere else.

"Do you think we can climb down into that?" he
asked, with consummate guile.

From that moment we were on our way.

Why the organization of expeditions demands more
frenzied work than launching a business or industry, I
have never been able to fathom. You need a valid
objective, finance, personnel, equipment, transport, logistics and so on for all three. Yet putting an expedition
together seems to run into many more snares and pitfalls

and makes greater demands on its planners. Of course, most of those involved are doing a full-time job as well and the product they are trying to sell, which is the expedition's objective *if it is reached*, is at best an intangible. Even so, take a look inside an expedition's headquarters and you'll see what I mean.

It is even more difficult in a country like Venezuela where such undertakings are not yet part of the culture. In the United States or Europe the explorer can take his projects to several established foundations which may or may not contribute both advice and funds. Here there is no such machinery, and although Charles had the encouragement and confidence of the Venezuelan Natural Sciences Society he met blank stares and incomprehension in so many of the contacts he made seeking funds and aid. But what he finally got off the ground after months of hard work was the biggest scientific expedition ever launched in South America, organized by Venezuelans, and carried through by Venezuelans and a few foreign specialists. Charles was a pioneer in more than one sense; he gave a start to the exploration of Venezuela by her own nationals.

Thanks to his hustle, the final list of sponsors was impressive: seven government agencies, the Venezuelan Air Force, the Organization of American States, four private institutions, and even one of the oil industry giants—Creole Petroleum Corporation, a subsidiary of Exxon.

Two points will give an idea of the logistics involved: The expedition personnel totaled thirty who had to be fed for a month with everything—down to the last grain of sugar—transported from Caracas. That makes roughly 2,700 meals. Charles made a list of meals in which no main item was repeated for three days. The list was followed every three days throughout the expedition and monotony successfully avoided. He then made a master list of every item of food required and in what quantities for 2,700 meals, calculated a sensible reserve and then stuck to it. Large expeditions can be very wasteful of food, increasing the quantities unnecessarily "just in case." There must be enough spare rations scattered round the Antarctic, the Himalayas and Greenland to launch scores of expeditions of young lads with no money to get started.

Getting the expedition into Camp One on the plateau needed a two-and-a-half-hour flight in a four-engine, Air Force C-130 transport to a scrappy airstrip near the Indian village of Cacurí on the Ventuari River, then a 35-mile helicopter shuttle to the Makiritare Indian settlement of Sta. María on the Erebato River, and from there a 50-mile shuttle to Sarisariñama. But not only men and supplies were carried by shuttle. The 50-gallon drums of gasoline had to be ferried by plane to Cacurí, then lifted to Sta. María and from there to Camp One by the chopper, for its own use. It was air transport which made the expedition possible for most of its members who could not afford the months that would be required to get to this objective by river and on foot. But this speed had one disadvantage over a long approach march. We three members of the team for the descent into The Pit did our best to get fit for the task. I personally gave up Scotch and even beer for three months. Toward the end I badgered the other two relentlessly to get them out on training climbs, but I finally had to give up. We worked long after midnight every night for the last two weeks. There simply was no time at all for keeping up to scratch physically. A long, arduous approach march would have put the edge back on our trim. As it was, we virtually went straight from long hours of overtime to the wall of The Pit.

For the others, who were not going down the hole, being within a pound or two of fighting weight was not so important, and most of them were, in any case, veterans of many expeditions in this type of terrain and knew how to move about. The doyen was William H. Phelps, Jr., a member of one of Venezuela's leading families, formerly a research associate of the Department of Ornithology of the American Museum of Natural History, and like his father before him, a dedicated naturalist and collector who has made the bird fauna of Venezuela the best known in South America. Billy Phelps is a big, craggy, blue-eyed man who looks uniquely what he is, a weather-beaten full-time explorer who has been probing the plateau since the thirties. With him was his Australian wife Kathy, extraordinarily lissome, indefatigable, a grandmother with the superb femininity to wear a chocolate-brown jersey sweater, tailored tan slacks and *beads* in the mud and ruin of Camp One. She was the outstanding personality of the expedition.

Julian Steyermark was the most renowned of our scientists. He is an American botanist, 15 years in Venezuela, who leads the world in the number of collections made. He is an impeccable expedition man who will keep his log come what may. With a red bandanna on his head, picking his way determinedly with his aluminum walking stick over the rough ground on the plateau, as long as there's light to see by, he is a heartwarming sight to lazy dogs like myself, who when there's no immediate action, sink into reprehensible sloth.

Another exemplary figure for the idle was G. C. K. "Stalky" Dunsterville, well into the second of two full careers. He is a dry, precise man, on the surface, but has a notably shrewd eye, as you might expect from a former president of a giant oil corporation, Shell de Venezuela. Having got to the top in that hard world he became a leading orchid specialist and collector who can match with his brush the fine art photography in the books he has published. Nora, his wife, is a tiny, sweet-faced woman, with a wit as mordant as his own and an expedition record as long as your arm.

The two specialists from the Organization of American States' scientific division in Washington were Braulio Orejas-Miranda, a Uruguayan biologist, and the Venezuelan geologist Antonio Quesada Estévez. Braulio never for a moment lost his good humor, he had an infinite repertoire of stories, and at night, after a hard day, when everyone was lolling in their hammocks, he would go off with his collecting bags, net, light rifle and flashlight to hunt down the snakes, lizards, frogs and what not, that are his speciality. Antonio, with his bandit's eyes and extraordinary knowledge of firearms, had been a mineral prospector, and as his beard grew longer and his baggy olive green pants grew muddier he looked as though his proper era would have been the lawless days of the California Gold Rush. He briefed us as carefully as Julian on what specimens we should bring up from The Pit.

Pablo Colvee was a geologist too, and a man to be envied. He works for Codesur, the Commission for the Development of the South, a government agency responsible for the pioneering of the thousands of square miles of the Territorio Amazonas, some of the wildest and most beautiful country on earth. For centuries the Indian tribes have traversed the region by its hundreds of rivers, the only way possible, living their own simple, unencumbered culture. Codesur is building heliports and airstrips there and driving dirt roads through the jungle. Meanwhile, men like Colvee and his team are roving the area studying the genesis of its formation, its geological history, and presenting the scientific world with evidence to back their new theories on the story of the Guiana Shield. He is one of a fortunate team pursuing, not the narrow interests of a mining company, but the opening up of a new frontier for Venezuela.

These men were the hub of the expedition. They would investigate the surface of Sarisariñama while we were down The Pit, and then move on to the neighboring plateaus, Jaua, to the northwest, and Guanacoco, to the northeast.

Charles, Jimmy, and I, and other plain workmen of the expedition slaved through the last night in Caracas packing stores and equipment, and running through the endless checklists. Over in a corner of the vast main floor of the Natural Sciences building, I was stealthily reducing our loads for The Pit from seven duffel bags to five. I say stealthily, because over the years my mania for light travel and spartan gear has grown so chronic that few fellow trekkers will fall in with it. Maybe they haven't spent so many days on the end hauling packs up rock walls. Maybe they are all nuts and have a fancy for this form of torture. At any rate, I surreptitiously heaved out my good friends' spare shirts and socks, cooking pans and tins of food, reduced the exaggerated collection of pitons and snaplinks, disposed of somebody's tennis shoes (slippers, for heaven's sake?), some spare batteries for the caving headlamps, refills for the carbide lamps, and hid them in an old sack that would not be going with us.

Then I tied up my masterworks of packing with devious knots so that nobody could nose about in them and start nagging, and at 4:00 A.M., dragged them outside to the equipment truck. By this time the scientists had arrived and we piled into a hired luxury tourist bus for the 90-minute drive to the Palo Negro air force base at Maracay, east of Caracas.

There on the tarmac stood the huge gray C-130, a barrigón or "Big-Belly," as they call it here, with its huge rear door let down to form a ramp. Under the military eye of the crewmen we labored on, humping our couple of tons of gear from the truck into its complicated innards, full of electric motors and winches, miles of cables, and the paratroopers' static lines of fond memory to several of us in the group. As the monster trundled off to the runway, somebody passed around a roll of toilet paper for people to plug their ears against the takeoff blast of the four engines.

It was 8:30 A.M. as we turned southeast for Cacurí and 11:10 when the pilot began his precisely calculated run-in to the airstrip. I had time to study that airstrip later—too much time, as it turned out. It is on high, open savannah, and its surface is hard, rutted earth, criss-crossed and dotted with lines and clumps of tall, indestructible grass. But that's the least of its difficulties. What matters is that it isn't flat. It rises to a distinct hump, then drops sharply away to be swallowed all too quickly by the savannah. Having no idea what it looked like as we came in we thought the pilot had hit the wrong button, for the instant he touched down the engines slammed into reverse with a din and vibration that had us all clutching in fright at our canvas seats. I'll swear the damned thing all but stopped dead, as it had to to avoid tipping up on its bulbous black nose and somersaulting into the hummocks at the end of the strip. I read in an airman's book once that Latins make good pilots. In the years I've lived here I've searched for that reference, without success. But in my experience it is right. When they took off, Lieutenants Rómulo Martínez and Carlos Silva revved the engines with the brakes on until the huge plane was visibly lurching and shaking. Then it shot forward up the hill, over the hump and vanished down the other side! What seemed minutes later it appeared again clawing slowly upwards and away. A superb performance.

As the plane flew off I walked a couple hundred yards across the savannah out of earshot of the others. To the west I could see the Indian village about a mile away. There was a line of trees curving northward marking the course of the Ventuari River. Further west a small plateau rose abruptly, its red and yellow rock walls bright above the forest. To the north, south and east, the savannah stretched for miles until the jungle began again, rising up over rolling hills. The only discordant feature was a dozen or so blue-painted 50-gallon gasoline drums about a hundred yards beyond the airstrip. I stood waiting for something I had thought of often over the past few months. A slight intermittent breeze rustled softly in the dry grass and a fly buzzed faintly nearby. Then the wind dropped and the insect was still. And there it was. Total, utter silence. If you are a city-dweller, try to recall when this last happened to you; when there was not only no noise of men and their gadgets, but none of animals or of nature itself. I peered at a cloud, ears pricked to see if it creaked a little as it moved. But there was nothing. This silence was absolute. I waited a little

longer testing yet again, the old legend that in these
conditions you can hear the world turning.

But I heard nothing. And as for the music of the
spheres way beyond, that we've read so much about, I
drew another blank, try as I might. I fear the truth is that
total silence makes your skull ring, until you acclimatize
to it, and that is what the mystics hear.

There was no chance for further investigation, for a
faint yell floated over to me and I turned and headed
back to rejoin the group. The workers were at it again.
We had unloaded the plane at the double and our gear
was in two jumbled piles on the airstrip. We lugged it all
off to one side and sorted out the items to go in the first
helicopter shuttle to Sta. María, the Makiritare Indian
settlement on the Erebato River, 35 miles to the east.

By now about fifty Indians from the village had
gathered to watch the show. The women wore dresses,
most of them with a fat baby astride one hip. The men
were in old slacks and shirts except for a few elders who
were naked except for the traditional guayuco, or
G-string, made of red cloth. There was one big fellow of
about fifty, with shoes and socks and a baseball cap. I
find this headgear is ludicrous on anyone except a fully
uniformed baseball player. But nothing could make this
man look silly. He was the chief, his name was Isaias, and
he had the honorary rank of captain normally given to
the headmen of villages. I went over to pay my respects
and was astonished to find myself staring into a Maya
face. Here were the same strong features I had seen in the
carvings in the ancient Mayan temples of Central
America. Those sculptures had always given me the
impression that the Mayas were squat, morose and

bloody men, and he had the same look about him. But, when I shook his hand and greeted him: "Buenos días, capitán," I found him lively and agreeable, and he welcomed me with easy expansiveness, as well he might, being boss of this whole vast savannah.

There was a sudden commotion among the Indians and then they all stood still and silent, facing east. For one bewildered moment I thought the long arm of Mohammed had reached them. Then Charles said: "It's the helicopter." But he couldn't hear it or see it. None of us could. Yet the Indians were right. It was coming. As the day wore on and the shuttles went back and forth, it became a game. We found that the sharpsighted among us could vie with the Indians in picking out the tiny dot of the craft as it came over the hills to the east, but none of us ever heard it before they did. The whirlybird clattered over our heads and beyond us to land next to the gasoline drums. I glanced at Charles.

"Fathead," I said.

"Dumb fink," he said.

Of course it had to land there; that's where it refueled. And we had a couple tons of gear neatly stowed under two huge tarpaulins, one hundred yards away from it. As penance, we grabbed two heavy containers each for the first shuttle and stumbled over to meet the crew. First was Major Pablo Martínez, a strapping, strikingly handsome man. Clark Gable with a deep tan and a big mustache. Next was Captain Andrés Gutiérrez, his trim, neat copilot. Then José Omar Uribe, the archetypal Master Sergeant Technician, fully aware that if it wasn't for him, and the likes of him, the Air Force would never get off the ground. They were a marvelously competent trio but utterly unpretentious and easygoing—except where José

Omar and the maintenance of his aircraft were concerned.

Pablo said he could make the round trip to Sta. María de Erebato and back in an hour and a half, including loading and unloading. He could take 1,300 pounds of kit and four people at a time. It was already noon, so we heaved in the first load—each sack and container had its weight painted on it and was suspiciously eyed by José Omar as it went aboard—the first four passengers jumped in, and off they went. Phase Two had begun.

There were sixteen of us left. We sorted out the personal kits of the next four to leave and made up the weight the helicopter could take with ration packs, and took it over to the pickup point. Then began the move of the main bulk of the stores. It was now that we began to feel the bite of the sun. Heat in the ordinary sense of hot, heavy air was not the problem, even though we were only four degrees from the equator. We were on the uplands after all, and there was a slight breeze most of the time. But there was no shade. The nearest tree was half a mile away, and our world was one big bright glare. They have the right word for it in Spanish. "El sol pica," my companions would say. The sun "stings." Soon the heavy work and the stinger began to show who was who among us. There'd be one man with a 50-pound pack on one shoulder and another under his arm. There'd be another groaning under the weight of a 10-pound ground cloth. The Indians joined in now and then, but lost interest after one trip. We were still at it when the helicopter came back, loaded and took off. Now we were twelve. We had half a five-gallon jerry can of water left and I found someone splashing it into his hand and wetting his head and neck, spilling it on the ground.

"What the hell d'you think you're doing?"

He stared at me with total innocence. "Just cooling off," he said.

"That water is for drinking. Not even that. For sipping."

The light of understanding shone in his brown eyes. "*Sí, hombre.* Of course."

"You know what you are?" I said, solemnly. "You're *improvident.*"

Jimmy, standing nearby, hooted with laughter. "Come on, *maldito inglés* . . . you bloody Britisher. Let's go fill these jerry cans in the river."

We picked up a couple of five-gallon cans each. Gene de Bellard-Pietri, the speleologist, who on his own admission was off form, had been humping loads like an ox. He came with us with a ten-gallon can. You fill that with water and you have 100 pounds to carry. The river was a little over a mile away. The trail to it led to two daub and wattle, palm-thatched huts isolated from the village. There was an Indian family there, a few hens and a starving dog. The river was 200 yards wide and ran quite powerfully in a long, slow bend. Here was the typical "black water" of all the rivers of this region. In fact, it is the color of champagne. In the shallows it is dark red, perhaps because of the pink sand beneath, and it shades into gunmetal, slaty black as it deepens. It reflects the trees on the banks in a startlingly true image. One glance at it and we had but one thought in common. A swim. We slid down the bank to the water's edge and whipped off our shirts. Then Gene stopped us.

"What about the ladies?" he said, nodding up the bank to the watching Indian women.

"They are Makiritare. That means you can show your ass but not your johnny wobbler," said Jimmy, or words to that effect. It was a piece of lore he had picked up from his brother Charles who had spent many months with the Indians and could even manage their language. For them, as it has to be if all you wear is a guayuco, naked male buttocks are normal, but naked male genitals are an affront. We stripped off to our shorts and dived in, drinking the pure water as we splashed happily around. The sting of the sun was instantly cured but it caught us again all too quickly as we lugged the jerry cans back to the thirsty mob at the loading site.

"Well, at least we've seen the Ventuari," said Gene, struggling along with his 100-pound load. As things turned out it was far from our last look at the noble river; we were to become commuters to it.

We found that some of the thirsty mob had decided they were in a bad way. Two or three had crawled under the tarpaulins out of the sun, exchanging a brazier for an oven, and were whispering hoarsely about their approaching deaths. Others were still plodding up and down shifting the gear. We joined them and at last everything was in order next to the gasoline drums. Each item was marked with a color as well as with its weight. Yellow for Sarisariñama, blue for the Jaua, red for the Guanacoco, the colors of the Venezuelan flag. Except for the OAS pair, Braulio and Antonio, all the scientists had gone in the first flights. We now had to decide who would be the next four to escape the grilling sun. It was nearing 3:00 P.M. and the breeze had failed us. We waited in silence for the helicopter, straining our ears to see if we could beat the Indians to it. Not a chance. They were chattering and pointing to the east long moments before we could truthfully say that we could hear even the suggestion of the sound of its motors. To our infinite regret we neglected to time the interval between their picking up the sound and ours. But this time we scored on sighting it.

The expedition doctor, Jesús "Chucho" Díaz, let out a whoop: "*Allá, Allá;* . . . there. Over the hill on the right." Sure enough we all eventually picked out the tiny dot at what looked to be an eighth of an inch above the horizon. Some of the Indians patted him on the shoulders smiling and nodding.

"*Muy bien, muy bien* . . . very good," they said.

Those who spoke only Yekuana, the Makiritare language, said something to him too, and Chucho, never at a loss, shook his head diffidently and, pointing at their ears, threw up his hands in astonished admiration.

Experts by now in the art of shuttle service, we were waiting in two groups on each side of the landing spot when the helicopter arrived. The instant it touched down we ran forward, ducking instinctively under the blast of the rotors, ripped open the doors, and slung in the prescribed weight of kit. The lucky four we'd chosen jumped in, chattering and laughing with relief to be escaping from the griddle. We slammed the sliding doors shut, gave a thumbs-up sign to the pilot and away they went.

Now we were eight. Except for the odd chocolate bar, an apple or whatever, which we had brought for the plane trip down from Caracas, nobody had eaten since the day before. After a bit of a wrangle we decided to keep it that way rather than interfere with the carefully calculated ration packs. We would all be in Sta. María de Erebato by the end of the day, anyway. Jimmy and I decided to go to the river again and on the way we marveled at the precision of the operation so far. The helicopter was returning every 90 minutes on the dot. It was just after 3:00 P.M. The helicopter should be back at 4:30 to take out another four men. It would then return at six, which would leave just enough daylight for the half-hour flight to Sta. María taking the last four men and the remaining kit.

Just then a lizard skittered across the trail ahead of us and we went after it like a pair of terriers. Jimmy got it with his cap. We were about a quarter of a mile from the dump and we turned back towards it holding our trophy on high and yelling for Braulio.

"Un largarto . . . we've got a lizard."

When he finally caught the word he came bounding across the grass. Marvelous fellows, these scientists, and fortunate too. Here was this specialist, forty-one years old, a man of the world, and a Washington sophisticate, grilled red by the sun, leaping across the grassy hummocks—for a three-inch lizard.

"Where is it? Let me see it. Don't drop it. Give it here . . ."

It was the first specimen of the expedition and he carried it briskly back to his herpetologist's box of tricks. He had work to do. Jimmy and I, a dentist and a writer, with not a smattering between us of the naturalist's arts, walked on to the river, knowing that there were a hundred things happening every square yard of the way, on the ground, in the grass, in the air, and we hadn't the craft to see a single one of them.

We got back to the dump in time to join the game again for the 4:30 P.M. shuttle, and again we were ready for a bit of split-second teamwork on each side of the landing site. We crouched in the blast as the chopper touched down and rushed for the doors. But this time the engines were switched off and the rotor blades slowed to a rhythmic waff-waff as they circled overhead. The crew climbed out stiffly and stretched themselves. We looked at them inquiringly. Major Martínez cocked his thumb at the gasoline drums.

"We have to refuel," he said.

As one man, we looked at our watches.

"How are things going over there?" Jimmy asked.

The major shrugged. "There's a bit of a hubbub. Charles says he wants the radio men in the next batch."

Lionel Jugo, an architect, and Federico Isaias, a student, like the rest of us with their burned faces turned away from the sun, beamed derisively at us. I had only met these two the night before and knew them only as radio buffs. Suspicious of any complexities on an expedition, I was indifferent to their role in the team. Later I was to admire thoroughly their know-how and finally to become very grateful to them. Had it not been for their varied skills, I might not be here to tell this story. Lionel and Federico stowed their gear aboard, and the debate began about who should go with them.

While José Omar was hand-pumping the gas into the chopper we stood aside with the major for a smoke. I pointed at my peeling nose. "We'll all be very happy to get off this hot-plate savannah," I said.

Pablo looked uncomfortable. He nodded over to the east. "You see how the cloud base is lowering over those hills?"

"So?"

"So this is the last flight today. Four of you will have to stay behind."

Dead silence. We stared at each other calculatingly.

Then someone asked Pablo what time he would come for us the next day.

"When the cloud lifts in the morning," he said. "If it lifts at all, that is."

More silence. "Well, I think I'll try a night out on the prairie," I announced.

Jimmy, Braulio, Antonio and Chucho chimed in at once. They were staying too. That left Armando Alicandu, at twenty-two, the youngest member of the team and nicknamed "Archie" after the comic-strip character. He was cook and general handyman, a sunny-tempered redhead, glad to undertake any task just to be along with the expedition.

"They *need* you, Archie," we told him as we propelled him towards the chopper. We loaded two extra containers equivalent to the weight of a man, slid the doors closed, and with lewd gestures waved them off.

Braulio drew his long knife-cum-saw-cum-kukri and grabbed a ration sack. This time no one demurred. In moments we had a goat cheese split into five portions, a tin of Spam opened, and were passing round the hardtack biscuits. This haute cuisine seemed to fan nostalgia in my companions for further creature comforts. They began to mutter about tents.

"Tents? You've got the whole wide savannah to sleep on," I said. "And if it rains, which it won't, you can spread a tarpaulin over the gasoline drums and crawl under there. What's more, this very night, February fifth, the moon is full. If you don't sleep out in such a place as this, and on such a night as it's going to be, you'll regret it forever."

But, rosy and gold in the noble sunset, this quartet was already burrowing about in the gear looking for the tent packs. As darkness fell they were still pitching their homes for the night, two four-man tents between them. One red, one green.

An hour later, dead tired, I was stretched out on my sleeping bag, star-watching, and hoping Braulio would emerge and offer me a cigarette. The day before, I'd stopped smoking for the duration of the expedition and hadn't brought any. My condition was already critical. Braulio did in fact emerge, but not for a smoke. He was carrying a net, some plastic bags, a headlamp and God knows what else.

"Now what?" I said.

He explained agreeably enough that he was going over to the river because certain nocturnal creatures would be up and about on its banks, and that not everyone felt that expedition time and money should be wasted by lazying about.

"Anyone coming?" he shouted suddenly, to round off his homily.

After a moment, Antonio, flashlight in hand, emerged from the green tent, while from its red twin we could distinctly register vibes of sheer disbelief. The dauntless pair marched off northwards and as their voices receded, I went back to searching out what constellations I knew. But with a full moon as lambent as this one, the stars were dim. I was still awake at midnight when the OAS hunters returned. Total bag: 19 tadpoles and one frog.

Before long the green tent was alive with snores and I picked up my sleeping bag and moved over to the middle of the airstrip. Braulio told me next day that he came out to take a leak some time after midnight and noticed that I had vanished. Groggy with sleep, he debated his next move. He said, deadpan, he finally decided not to call the headman because an extraterrestrial snatch was outside the capitán's jurisdiction. Braulio will tell you anything.

The sun got us up shortly after 6:00 A.M. and was soon frying us again. We got the tents down smartly, packed everything up and sat staring eagerly towards the east. Chucho swept his arm around that arc of the horizon: "You see? There's no cloud. Pablo should be here to pick us up any minute."

Any minute? It was going to be ten hours.

The any-minute-now theme ruled our actions the whole day. We felt that if we strayed far from the landing site looking for shade the helicopter would choose that moment to arrive and we would hold up the expedition schedule. After four hours under the grill we noticed that groups of Indians were converging on the airstrip along several trails from the north and west. Eventually Isaias appeared and was astonished to find us still there. He shook his head and waved his hands in reproof. Why had we spent the night there? Why hadn't we come to the village to be his guests? Why hadn't we availed ourselves of his ready hospitality? We tumbled over each other with excuses and explanations, pained to see how genuinely upset he was. We were, in truth, guilty of a real breach of protocol and kicked ourselves for not having understood this the moment we learned we had to stay behind the day before.

We still hadn't understood what the gathering was all about until the Indians took up their familiar aircraft-detection stance. This time, however, they were facing north. We soon picked out the familiar black dot on the horizon but as it approached we saw that it was not a helicopter but a C-130. The great, fat plane roared over our heads, made a wide, slow circle and then came in. There really is nothing quite like one of these monsters landing on a short, rough strip. Almost at the instant the wheels touch down, the whole forward impetus of the great machine is resisted by the reversed power of the four engines. Clouds of dust, pellets of earth, and clumps of grass billow up fore and aft. It is a scene of pure violence and cacophony. And when all is still it comes almost as a shock to see how minute and fragile in comparison are the men who control all this elemental force, as they finally emerge from the vast gray hulk.

By now, Isaias and his elders, with all his people behind them, were waiting silently at the rear of the plane while the ramp was lowered. We sensed there was something special about the occasion so we remained discreetly to one side. A group of civilians came down the ramp, papers and folders in hand, and obviously filled, themselves, with some sense of impending ceremony. Isaias stepped forward with great dignity to greet the visitors. And then in that greatly endearing Venezuelan way, the formality was swept aside and everyone was exchanging the traditional abrazo—a gesture in which you face your man, throw your arms round him and thump him warmly on the back. Now everyone formed an expectant circle and there was a strange rumbling from inside the aircraft. And then down the ramp and into that pure, green, limpid world there chugged a blood-red tractor, its blackened exhaust spewing evil, chemical-blue puffs into the crystal air. While the Indians clapped in joy and pride, Braulio and I stared at each other, speechless. I watched the play of emotions on his face—astonishment, indignation, and, finally, what seemed suspiciously like grief. A bright young man came over to us. He was from the Ministry of Agriculture. This was the beginning of a new experiment in cultivation, he explained. It would widen the Indians' potential and might even bring them into the market. We nodded in resigned understanding. "Commendable, very commendable," Braulio muttered. But we knew that we had been witness to the beginning of something else; of the inevitable cycle—mechanization of agriculture, commerce, industry, pollution, alienation. Progress.

About an hour later the great, gray airborne monster successfully made its hazardous takeoff, circled, and was gone. The red land-borne monster chugged off to the north with its attendant crowd of happy Indians, and we were left alone. It was noon. The sun, straight overhead, was pressing on our skulls, and under our shirts the skin of our shoulders was stinging again.

About 2:00 P.M. we unenthusiastically ate the remains of the goat cheese with a sardine each from a warm tin. Then after a long look to the east we gave up and retreated to the nearest trees. It was a swamp rather than a forest and we squelched our way in to sit on hummocks of earth surrounded by stagnant pools of oily water.

We each cut a two-foot, rubbery leaf from an odd-looking plant and fanned ourselves in the blessed shade. You get some good conversations going at times like this, when strangers are thrown together in improbable circumstances. Suppose some observer happened to find us there; a lost forestry commissioner, or whatever. What intricate story could we give about why five men, from London, Caracas, Washington and Montevideo, were sitting in a remote swamp, fanning themselves with leaves at 1430 hours, February 6, 1974? But we didn't get very far with our inventive theme for we were soon crawling with mosquitoes and *bachacos*, a half-inch ant with a bite as hideous as its big-jawed head.

We returned disconsolately to the dump, and Jimmy and I, quite irresponsible by now, went off to swim in the river, leaving the others sprawled under a tarpaulin stretched across the gas drums. We were about to set a world record.

When we got back we found three or four Indians had joined our friends in their makeshift oven to get out of the sun. They had brought us a few lengths of sugar cane and a pineapple, which Braulio was dividing with surgical precision. A few yards away Isaias was standing, staring into the distance, hands behind his back, legs astride, moving rhythmically up and down on the balls of his feet. There was something very familiar about this stance. Squatting in front of him was the village teacher, a Spanish-speaking Indian, pencil in hand and writing pad on his knee. Isaias, so help me, was dictating a letter. His pose fell into place now. He'd got a tractor in the morning and here he was staring out over Wall Street composing a memo to the chairman of the board. The teacher folded the message and handed it to Jimmy while Isaias explained he wanted it taken to Sta. María and delivered to his fellow chief, Capitán Pedro.

We were still digesting this episode in evolution when Chucho whispered: "Look at the Indians."

It was 4:30 P.M. They were facing east listening intently. That was enough for us. We rolled up the tarpaulin and divided the gear into two piles each side of the landing site, and went over to say good-bye to the chief.

He nodded towards the east. "It would take weeks, maybe months to get over to Sta. María without a helicopter. When you get there you will be the first men ever to swim in the Ventuari and the Erebato rivers in the same day."

East Africa. Their attitude is nearer to kindly . . . good grey

3

You would have to go a long way in the history of master races to find so superbly arrogant and so succinctly expressed a credo as that of the Makiritare. It is this: Only Ourselves are People. In short, everyone else, including you, is an animal or a slave. But they are not jackboot oppressors of neighboring tribes, nor do they stare a stranger haughtily down like, say, the Masai of East Africa. Their attitude is nearer to kindly condescension. Notably serene individuals, their elders will watch the incomprehensible antics of their rare white visitors with grave attention but will not demean their own dignity or offend yours with even the ghost of a grin. Neither will they admit to being impressed. Show them a chainsaw, a shotgun, or a walkie-talkie, dream possessions all, and they will not for an instant slacken their composure. They are the grandees of the jungle.

There are only about 1,400 of them spread in about 30

settlements over 18,000 square miles of territory. Their forbears survived the massacres of the Spanish conquistadors, who killed off more than 90 percent of the Indians of South America, because they were able to retreat up into the headwater region of the great rivers—the Caura, the Ventuari and the Erebato. No large contingent of murderous colonists could follow them because of the rapids and the thick jungle. Over the past four centuries parties of Indians emerged from their fastness to trade. They picked up iron tools from the whitemen, but their base remained free of contact with them virtually until the first permanent penetration in the 1950s by missionaries of various religious orders.

They are forest cultivators who cut and burn a clearing, grow yucca or other root crops, and when the plot is worked out, abandon it and clear another. This sort of jungle gardening began about 5,000 years ago and the Makiritare are one of only four tribes in Venezuela where so ancient a pattern of life may still be observed.

The first lesson their culture has for our own is immediately apparent to the visitor. The Indian is unemployed, by instinct. He is not jobless, because "jobs" don't exist. The out-of-work on the street corners of our own cities are unfortunates. The Indian sitting on a log is in his natural state. He can go off and clear a plot for the women to plant, work on his boat, weave a basket, or he can stay put. The decision involves his needs and the way he feels. It does not involve guilt, the work ethic, or the inability to be idle with grace. The simplest human community has its own built-in pressures, of course, but I wish mine were of the minimal variety supported by the Makiritare.

When their respected chief Calomera died in 1959 the Indians burned their communal dwelling on the banks of

the Upper Erebato and moved downstream to found a
new settlement now called Sta. María de Erebato. About
this time, missionaries from the French congregation,
Charles de Foucauld, had managed to get their presence
accepted, and this probably accounts for the sadly
un-Indian name for the village. Perhaps also under the
influence of the evangelizers and perhaps because of
what they had seen on their trading trips to far-off towns
on the Orinoco, the Indians abandoned the huge communal house to adopt the creole system of separate
houses for each family, set out along a 25-yard-wide
boulevard of hard, sun-baked earth. They did however
build a big communal dwelling for travelers or visitors
from other groups of the tribe.

Let's take a walk through this village. The first house
on the left has, like all the others, dried mud and wattle
walls supporting a thatched roof. It is painted white. At
the door three girls are talking. They are four-foot-nine
or -ten, with short basin-cropped black hair. Two are
wearing cotton dresses; the other is bare-breasted, and
pert breasts they are. (By the time we come back, she'll
have them covered. Missionaries again.) They have
strings of white and blue glass beads wound tightly round
their upper arms, below the knee and round the ankle.
Their eyebrows are completely plucked out, and their
faces and parts of the body are painted in dark brown,
blue and red, pigments which they boil down from seeds
and leaves, and mix with palm oil. The red mixture has
more than cosmetic significance, and is supposed to ward
off disease. They are chunky dollies, well-padded, like all
the Makiritare, but without curves. None is pretty, but
each has something that attracts, sloe eyes, a wide mouth,
a snub nose.

If your mind is on what it's usually on, you would have

to stay around for some time until your alien look has worn off a little. The girls have complete sexual freedom and for some unaccountable reason may be curious to find out what it's like with a maggot-white, midge-bitten dude from outside. But you won't get far by dangling glass beads and greenbacks. That bit is unknown among the Indians. As always, discretion comes first, and even established couples go into the forest for sex. However, if you try this, you'll discover you've got ten seconds flat before the ants are nipping at your ass.

Besides the next house is a *caney*, which is a shelter with four corner poles supporting the thatched roof. Here four older, less adorned and more businesslike women are dealing with the various stages of transforming bitter yucca, the staple Makiritare crop, into its various edible forms.

So important is bitter yucca to the tribe that it has its own legend. The hero is Cushu, a monkey who went up to the eighth and highest ring of heaven, home of Wanadi, Son of the Sun, stole the plant and brought it down to their ancestors. But it has an immediate drawback: its juice, mostly prussic acid, is poisonous, and must be extracted.

The women peel the yucca, grate it and stuff the bits into a long cylindrical press miraculously woven of bamboo strips. The press is anchored to a pole and then stretched by a lever until the juice runs out. The yucca comes out of the press in a solid column which is broken up and strained through a basket sieve. Most of the flour is being made into cassava, round cookies three feet across which are baked on an iron or clay slab over a fire and then laid on the roof above to dry. A smaller share is cooked in pellet form. It is called manioca, and you can toss a handful in your soup, or eat it with water.

Farther up the "street" you come across a cage of saplings at the side of a house. In it are four hunting dogs yelping and baying in an ecstasy of yearning to get out and tear you to pieces. Dogs like these have been locked up all over the village because of your visit and will stay that way until you go. The Makiritare are avid hunters, and sometimes the whole community will take off for weeks at a time, camp out, hunt by day and feast on their bag at night, until the area is cleaned out of game. They may use bows and curare-tipped arrows, blowpipes and sometimes ancient single-barrel shotguns, depending on the size of the prey. In the big league are tapir, alligators, deer and peccary. This last is a peculiarly ferocious wild pig whose tusks are prized as ornaments throughout the country.

At the end of the street on the edge of the airstrip is another breed of chaps, smaller and skinnier, with a great wad sticking out between their lips made up of tobacco and ashes. With their faces distorted by this protuberance, you cannot tell if they are grinning at you or not. A family of them have half a dozen hammocks slung in a *caney* there, black with filth, and in them swings everyone from grandfathers to a week-old baby. Beneath them are their smoking fires, chickens, scraps of food and the odd cooking pot. These are their total possessions. They are from the Sanema tribe, real primitives often described as slaves to the Makiritare. They look remarkably contented for people of that sad status, and you wonder how many western families could spend their day chattering happily together without Dad going off to the pub, and the youngsters clearing the hell out of it to wherever youngsters hang out these days. Obviously, slave is not quite the word. The Sanema will come to work for the Makiritare for a determined object, a tool or a length of cloth. When they have earned what they want, they disappear.

When we bundled out of the helicopter on the airstrip there were few Makiritare there to greet us. After the traffic of the previous day, we were no longer news. They waved us cordially on through the first huts towards the center of the village. But none offered a hand with the baggage. Like Guards' officers, no Makiritare gentleman would be seen dead with so much as a parcel under his arm. We struggled along looking for the rest of the expedition when we caught sight of the river down a 70-foot slope and only 100 yards away. We dropped our packs on the instant, scurried down the bank and stripped off. Isaias's prediction of our world record was about to be made fact.

Once again we knew that rare delight: to be hot and thirsty and to dive into a river so pure you can drink the water as you swim. The current was strong, and down-stream from us the surface broke and tore at the rocks of a minor rapids. The next Indian village down that way was anything from five days to two weeks away depend-ing on the season and the depth of the water over the rocks. Upstream was a lordly, sweeping bend where the river widened to about 300 yards across. Along both banks a rich, tall forest soared up and rolled away over the hills. The odd, melodic cry of a bird was the only sound, in this untrammeled beauty and peace. The only sound, that is, until a long, low dugout canoe came broadside on round the bend, out of control, at the mercy of the current, paddles flailing hopelessly. The stillness was rent by roars of anger and despair, and hysterical

laughter. Lionel, Gustavo Chami, the cameraman, and
soundman Hector Moreno, insouciantly, had gone for a
little row along the bank. But they had ventured into the
subtle world of the river Indian and the unseen forces of
the dark water had seized them irrevocably. They came
on helplessly, wildly rocking their craft which, with only
inches of freeboard anyway, was shipping water by the
gallon.

"Either they sink, or they'll go straight down the
rapids," said Braulio.

"Let's hope they sink," said Jimmy, in swift calculation
of the alternative evils. There was nothing we could do.

The dugout was now well down astern where Lionel
was sitting, hooting with laughter, his paddle held in
surrender across his knees. Gustavo rose shakily to his feet
in the rocking craft, threw his beloved leather hat toward
a sandy stretch on the inside of the bend, took a final
wild-eyed glance at the approaching rapids and flopped
overboard. Hector followed him and they both headed
for the sand with a desperate dog-paddle. As the boat
went under, Lionel, helpless with laughter, still sat there,
raising his arm in final salute as he vanished under the
surface. As we watched the subaqueous finale of the
comedians' act, we heard a sharp, rhythmic crack of
wood on wood behind us. An authentic river sound. We
looked downstream to see a dugout moving deftly up
through the rapids, swerving with immaculate judgment
from the leeway of one submerged rock to another, and
in each bold rush across the full current running between
them the beat of paddle poles against the hull rose to a
staccato drumming. The stern paddle flashed across the
hull to the offside, cut into the water at an exact angle
and the craft edged neatly into the still shallows under

the bank. We clapped and cheered in admiration and the crew looked across at us with shy smiles. There were two of them—Indian girls about nine years old.

Meanwhile Lionel, swimming alongside his water-logged boat, had brought it across to its mooring on the village side of the river. As he was bailing it out, an Indian elder appeared and shouted something in Yeku-ana. Charles who had now joined us, called to Lionel.

"Raise your paddle to show him it's not been lost." The Indian lifted his hand in acknowledgment and spoke again. The message, translated by Charles was from Capitán Pedro, the chief: Further tomfoolery with the boats was forbidden.

Rightly so. A dugout, called a curiara, is a superbly tooled artifact, a product of centuries of experience and skill. First, a mure or cachicamo tree of the appropriate girth is felled and trimmed. The trunk is dragged laboriously through the forest to a clearing. A strip of the outer wood is cut from one side, and starting from this the inside is slowly burned out. As it is hollowed, short lengths of timber are wedged across the opening until the curved line of the hull is achieved. A curiara may be a slim, one-man boat, or a 35-foot family affair with a 4-foot beam. There are no seams or joints. The whole fabric is one single piece and it may take a full year of careful craftsmanship to make. The paddles, too, are one piece, from the curved hand grip down the 3-foot shaft to the wide, heart-shaped blade. They last an Indian his lifetime, gradually polishing and wearing to the contours of his hands.

Journeys along the rivers may last for weeks down-stream and more on the return against the current. On tricky stretches one man stands in the bow with a long pole to keep the curiara off submerged rocks. The poles are also used to force up low-angled rapids, and when the current becomes too strong everyone goes overboard to wrestle the boat up, slipping and stumbling over the boulders on the riverbed. The Indians will shoot rapids downstream when the water is deep enough, and where it is not they carry the cargo overland to a point below the rapids and then return to manhandle the boat down the rocks. These days of course, outboard motors are appearing, clamped incongruously on the stern of these ancient craft. The purists may sigh, but for the Indian faced with a 100-mile shove upstream, it is a different story.

Life seemed a different story to us too for the moment, stretched out contentedly on a rock at the riverside with the sun sinking behind the hills. Then one by one we began to scratch, absently at first and then with growing concentration. Finally we were all sitting up examining ourselves for the source of this fierce itch. Archie had come down for his evening scrub and stripped down to his shorts. With the usual milk-white skin of a redhead he was a startling sight. He was covered up to the thighs with nasty, pale lumps each with a red spot in the center. Poor Archie, we thought, what dread ailment had hit him? He stared at us and looked down at his legs.

"You needn't look at me like that. You've all got them, even if you don't know it yet. They are chivacoas."

The chivacoa is a tiny red horror which burrows under your skin, raising a painful bump, and sits there for a couple of weeks making your life hell. There is no way of getting at it or squeezing it out once it is dug in, and there is no lotion, salve or ointment that will relieve the burning irritation. Archie had made his own investigation among expedition members and found that those who had done the most work unloading the helicopter on the airstrip were the worst afflicted. Therefore the chivacoas were concentrated in the long, dry grass at the landing site. Those wearing high boots, lacing up to the knee, were less affected. I recalled having slid down a half-mile of steep, dry grass below the sugar loaf mountains of San Juan de Los Morros years ago and ending up with 137 garapatas under my skin. They tormented me for five weeks. They are bigger than the chivacoa, but the bite is about the same. Within hours our own lumps began to appear. My score was 11 in the right armpit, 7 in the left, 13 across the stomach and 25 on each leg below the knee. But the medals went to the two or three of us who had one on their what's it, and were man enough to fork it out and prove it.

The expedition was quartered in a churuata in the center of the village. A churuata is a large building which houses several families. It is circular, made of 2-foot-thick mud walls about 6 feet high and surmounted by a 30-foot-tall conical thatched roof supported by an ingenious spiral system of poles which enable it to be built without scaffolding. It was about 50 feet across inside, with an earth floor and no partitions: one big, communal room. The scene inside was demented. The thirty expedition members had slung their hammocks every which way, dumped their baggage in a space between two long wooden tables, L-shaped, and were all frantically busy doing nothing that looked constructive. To one side the radio men had set up shop behind a rope barrier and were yelling a message to someone in Australia, by the sound of it, amidst the bleeps and crackles of their instruments.

Across the floor from them were two long benches on which sat a group of Makiritare, with their impenetrable faces, drinking in the scene. There was always a handful of Indians on these benches throughout the day, we discovered. But the big moment came with our nightly meal. This fascinated our hosts. They would begin arriving an hour before and by feeding time the benches were packed so tight that Capitán Pedro and his elders had only one arm free to handle their rough-rolled cigarettes. Throughout our main performance and in fact for most of the day, they spoke to each other very rarely and even then it was in short asides made without turning the head or taking their eyes off us. You have to watch your cigarettes, by the way, with the Indians, especially when your supply is part of your expedition baggage and therefore limited. If you offer one as a friendly gesture to whomever you are talking to, everyone near will point to the pack and then to his lips. So they each get their first and only manufactured cigarette in months, which they puff at inexpertly and wantonly, and instantly forget, and you're left with nothing of your daily ration. Their own are a mixture of coarse tobaccos, loosely encompassed in corncob leaves, and taste like burning tea bags.

Our kitchen was installed in an empty hut a hundred yards from the churuata. Large aluminum pots were set up on butane stoves and tended by Archie, Kathy and two Spanish-speaking Indians. Charles hovered about them demanding rigid obedience to his meticulous lists. The pots were carted to the churuata and spaced out on one of the tables. Thirty expeditionaries would queue up, and hold out their metal bowls for a handful of nuts, dried fruit, ear-shaped biscuits, or cookies called catalinas and made of molasses and maize flour, a piece of sweating cheese or limp Spam, and a mug of milk-powder goo laced with Ovomaltine or Semolina, I think. Afterwards, to console us all, came a brew of real Venezuelan coffee, strong and aromatic, and morale went up an instant ten points. Kathy, smart as paint, her unrepentantly Anglo-accented Spanish delighting every ear, presided as she does at a Diplomatic Corps ladies soiree at her Caracas mansion. But it was jungle boots instead of Gucci shoes and a dirt floor instead of marble. Above the table hissed butane gas lamps, pinging and cracking as the big moths collided with them. "Puedo ayudar a alguien . . . can I help anyone?" she'd call, as she circled the table to see that everyone had his ration of raisins. The mimics among us soon picked it up and Kathy's voice, offering tea or aid, could be heard floating from all corners, all day.

After supper Charles reminded the company about the rules of behavior for visitors to Indian settlements: if you swim you can show your naked behind if you think it's worth it, but not your old man. If you want to take photos of the Indians, ask permission first, and if they don't speak Spanish ask it by gesture; never enter any of the huts unless invited; always be respectful of Makiri-tare pride. "And now, if you look around you you'll see we have already turned their guest house into a pigsty. Why don't we start by picking up the cigarette butts?"

The Indians in the ten-dollar seats on the benches leaned forward in unison, maybe a quarter-inch. This was something that had to be objectively observed and scientifically recorded so that when the incident came up for analysis later there would be no doubts as to what had actually been seen. Among the expedition members were men of serious and sober mien, with many years of authority marked on their faces; and there were young scamps who punched and thumped each other and were eternally skylarking about. But all of them were now stooping to pick up butts and scraps of paper. Thirty rounded backs bobbing among the kit and the tables, with their labored breathing and one hand full of junk and the other scraping in the dirt for more. Finally everyone dumped his collection in a garbage bag nailed to a pole and the job was done. There was a common but almost imperceptible sigh from the benches and the line moved back that quarter-inch to the vertical again. Braulio, however, was about to break it up completely. First he laid out his kit on the table: flashlight, collecting bags, headlamps, folding rifle. The eyes of the Indians widened a fraction. They were great hunters themselves. Then he sought the Spanish-speaking liaison Indian and explained his needs. He wanted two men and a curiara to go upriver and stealthily probe any side creeks. The liaison man was intrigued and we could see the gleam in the eyes of the elders as he explained the proposition. But there whispered questions among them "Váquira . . . danta . . . lapa?" What animal was this sharpshooter planning to roust out and kill, at night and without a pack of hunting dogs? The liaison man came back to the table to clear the matter.

"Frogs," said Braulio.

The liaison man stiffened and peered closely at him, then turned back to the benches. There was a stricken silence for a few moments while Makiritare dignity fought with a monumental craving to scream and roll on the floor laughing. One elder turned weakly and laid his head on the shoulder of his neighbor, shaking softly. A few more, unable to contain themselves, left hurriedly, presumably to beat their heads on the wall in an agony of glee in the privacy of their huts. The chief nodded his permission and gave a few orders, and the hunt was on. I think Braulio and Antonio were a mile or so up the Erebato before the remaining elders had the strength and composure to rise from the benches and go home.

This delicious scene had gone unnoticed by the expedition team but Braulio, as aware of it as I was, took it all with great forebearance and good humor. This was not lost on the Indians by any means and the next day Braulio was to cap it with another, different performance that was to earn him a Makiritare accolade.

One thing we five lost cowboys had learned on the prairies of Cacurí was the joy of sleeping in an uncluttered world. We were now having to adjust rapidly to its antithesis, the nighttime bedlam of a churuata. Jimmy and the doctor shouldered their way into a festoon of hammocks to sling their own where not even a clothesline would fit. Those already established in that space, their territorial prerogatives threatened, were flailing away at them with sleeping bags and dirty socks. The riot rose in crescendo as the radio men ripped off their headphones in exasperation:

"Cállense la boca, coño de . . . shut up, you mothers!"

Farther along the wall a group centered round the pilots was roaring away as joke piled upon outrageous joke. Round a table of their own the film team were in violent argument over the packing of their tens of thousands of dollars' worth of equipment. Cutting incredibly through it all, were the astounding snores of an unknown Indian who was on his way God knows where upriver. He had tottered in, slung his hammock and fallen into it immediately. He must have paddled a thousand miles that day. Julian Steyermark, totally unperturbed, was deep in some botanical plot with his assistant, Victoriano Carreño. A stricken cry rose in some dark recess as Chucho lanced someone's festering elbow.

When the ladies and their spouses went off to the huts that had been allotted to them, our boys really broke loose. What on earth was being thought in the silent village around us I can't imagine. Indians seem to go to bed with the sunset and that was three hours ago. I had to think about my own bed too. I had no hammock. I found a tarpaulin folded on the floor, stamped to death the half-dozen mammoth cockroaches which had occupied it and rolled out my sleeping bag. I had been silent up to now, content that my part of the preparations had been accomplished with the five bags for The Pit meticulously packed, sealed and with snaplinks already attached. Then Charles came over with patently phony unconcern and drew me to the table.

"Packs for the hole all okay?" he asked airily.

I pointed to the five duffel bags standing upright, perfectly balanced, yellow tabs and snaplinks gleaming in the lamplight. He nodded. "Well don't forget this lot," he said, making swiftly for the door.

This lot was a 16mm Bolex movie camera, several lenses, ten cans of film, and a walkie-talkie radio in an 18-inch box. It couldn't have been worse. I'd rather deal with five dozen eggs and a chandelier. Out went all personal kit. We would spend our six days down The Pit in what we stood in. Out too came the helmets, the harnesses and all the climbing ironmongery. We'd have to make the shuttle to Camp One with all this stuff draped over our shoulders, like a trio of proper show-offs.

I cussed and fiddled and tinkered until, at midnight, the job was done. It was a tedious business, but it gave me the chance to see how the noise and bustle of an overcrowded churuata wound down for the night. And also a real bit of practical Indian magic.

There were twenty-six expeditionaries and six Indians, 32 hammocks in all. Conversations became desultory, with longer gaps. Then there were segments of silence within the dark circle. The botanists were asleep first. Then the helicopter crew. A few whispers fluttered here and there and then all was still. The empty moment had arrived that bridges the noise of the day with the sounds of the night. But it didn't last long. The unknown Indian shifted in his hammock and one of his monstrous snores rolled forth in majesty, reverberating across the hills southwards towards Brazil. His tuba was soon joined by lesser performers, the bassoonists and the buglers, and then the lowly mob of croakers. But he was too much, that Indian. Here and there people were stirring, blearily wondering what to do about him. Then one of his fellow tribesmen silenced him with the damnedest trick. He just blew him a kiss. The loudest, wettest, smackiest kiss. The stuff of the nightmares of small boys with fat, effusive, damp-lipped aunts. Then another smack, and then a stream of about half a dozen. The snores stopped. Many a man lay awake after that performance, pondering it in wonder, and like me, itching for a chance to try it. We didn't wait long. The tuba let go again and from all over the churuata there burst a chorus of fat aunts in full smooch. Then gurgles of laughter and silence again. I picked my way through the hammocks out of that stifling hut and lay on a bench outside. These Makiritare had the fine-honed senses we must all have had once upon a time, I thought. The next day we were to see these senses form a vital link in a chain that was otherwise pure, hard science. A chain that had three lives dependent on it.

We were up at dawn, fed in minutes, and standing by to get out of that hot haunt of chivacoa and cockroach. Outside the churuata an eager group was gathered round the pilots who were sniffing the air and gauging the cloud ceiling. Charles was putting the pressure on.

"Come on, Pablo. Let's risk it. Let's just get the first group away." The first group was himself, cameraman Gustavo Chami, and two Indians he had hired the night before. They were to land on Sarisariñama about a mile from The Pit where aerial photos showed what appeared to be a clearing in the jungle with a hard surface. They would level this landing site, clear a zone for a supply dump and make a start on Camp One itself. The next group would be Antonio with Pablo Colvee and his Indians who would prepare Camp Two on the edge of The Pit. The scientists would follow. Camp One was 50 miles away, but that was the least of the problems. To get to Sarisariñama the pilots had to find their way across the intervening plateau of Jaua, whose gigantic, wildly crenellated west wall was a formidable obstacle in the clouds.

Pablo looked at his copilot Andrés, who nodded.

"Bien. Vámonos . . . let's go."

Three or four of us grabbed the first load of gear and took it through the village to the airstrip. At its edge we paused, eyeing the long dry grass. Chivacoa territory.

Jimmy shrugged. "There's no way round it, and anyway we're half-eaten already. Come on," he said.

We loaded up the helicopter, gave our vanguard lads an envious slap on the back and off they went in a fast rising slant to vanish in the mist. Phase Three was under way.

We trudged back through the village to ferry the next loads over and returned to fidget around, scratching, in the churuata. About an hour and a quarter later the radio crackled into life. We heard Lionel talking with some urgency into the transmitter, and then the voice of the pilot, smothered in static. But the tension came through clear enough. Lionel glanced briefly up at us.

"They're lost," he said.

He returned to the transmitter. "What can we do, Pablo? Just say what you want."

There was a rapid-fire, crackling answer, in those blurred metallic tones only the practiced radioman can understand. Lionel jumped up, pulled the cables off the radio and rushed it outside.

Federico flung the cables out of the window, ran through the door, picked them up and led them over to Lionel who was shouting at us to bring a compass and collect some Indians. Compasses and Indians? Now we could hear the helicopter somewhere up in the clouds. The pilot had found his way back but couldn't see the ground, let alone pick out the village or the airstrip, and with the abrupt hills surrounding us there was no room for the slightest mistake in his approach.

The helicopter had a radio direction finder but Lionel had no equipment for it to beam onto. The human-scientific chain which was being rigged to fill this gap looked lunatic. But the pilots knew their Indians. Federico, compass in hand, stood behind Lionel with a little group of Makiritare. We listened intently to the noise of the helicopter's motors swelling and receding above us in the mist. I tried my damnedest but couldn't tell whether it was north or south or where the hell.

Federico nudged the Indians. "Where is it?" he shouted.

Instantly and without even a second's consultation between them, they pointed together at some vague spot in the gray pall. Federico swung his compass in line with their outstretched arms.

"One-twenty degrees," he said.

Lionel transmitted this bearing and we could make out Pablo talking to his copilot.

"Okay. Let's go in at three hundred degrees."

The clatter of the motors grew rapidly louder and the tension increased by the second. I pictured the pilots staring through the windows at their feet, side-slipping desperately as some tree-clustered hilltop flashed at them through the mist. We could hear the sudden roar and dip of spasmodic accelerations, a lurching, ugly sound of tortured engines.

And the steady voices of the radio men: "Ninety degrees . . . hundred ten." And the Indians, heads cocked, slowly swinging their arms in an eastward arc following the path of the approaching craft.

All at once it shot into view, trailing cloud like spindrift, not a hundred feet above our heads, its bright red and white stark against the gray. We jumped and cheered, hugging the Indians and thumping the radio wizards on the back. Then the whole crowd was stampeding through the village to the airstrip to find the crew climbing slowly out while the rotors slowed and the whining turbine lowered its pitch to a soothing hum.

The pilots lit cigarettes, put their heads back, exhaling a big sigh of smoke at the clouds that had trapped them.

"Well, Pablo? Andrés?" we burbled.

Pablo put his arm round the shoulders of an Indian, and waved a nonchalant hand.

"Nothing to it. With these guys around, who needs radar?" he said.

That performance put the Indians so far ahead of us in the game of one-upmanship that we really thought we'd lost for good. Our sharp-eared friends were well aware of their lead and their wide grins were beginning to have a sly twist to them. But they hadn't won yet. Braulio was about to step onstage.

We were walking back towards the churuata thinking we would have to find someone among us who could swallow a sword or tear a telephone book in two, when we saw him emerging from the trees onto the flat, baked earth on the edge of the village. Those sly grins grew even wider. Here was the tadpole hunter. Braulio had his light rifle in one hand and a collecting bag in the other. As we came up to him a lizard zipped from a tuft of grass towards a heap of stones. It moved as fast as the eye. But Braulio was faster. I'm sure the rifle cracked before the collecting bag hit the ground. The lizard was neatly drilled through the neck. We stared at the corpse, and at Braulio, and then at the Indians. And slowly, widely and slyly we grinned at them. They wagged their heads and whistled and acknowledge the point. But Braulio hadn't finished yet.

When we reached the churuata he fetched out his trick-box. It was eighteen inches high, two feet wide, and round. He opened his collecting bag, inside of which were several plastic bags, in which were *fourteen toads*. Monster toads. *Bufo marinus,* they're called, and the name does them no justice at all. One of them, a female, was nine inches high sitting on her haunches, with a head three inches across. She weighed four pounds and is one of the most poisonous creatures in nature. On her back, clutching her tightly was an apple-sized male. He was in the coupling position but he looked more terrified than randy to me.

I picked up a bag in which five of them were squirming around and made a new discovery: toads are *strong*. If they come any bigger than that first female, I'm sure they can knock a dog down.

In another bag were three frogs. *Leptodactylus ocellatus*. The edible frog whose hind legs delight the gourmet. These were big too, bigger than any you'd find in a French restaurant. Braulio held one up to the Indians.

"You like these? Good to eat, eh," and gestured towards his mouth. Their spokesman drew himself up a little and waved a disdainful hand at two Sanema who stood nearby.

"*They* might eat such things. *We* do not."

It was a slightly uncomfortable moment but it passed the instant our herpetologist opened his box. We crowded round as he took out his kit—bottles, flasks, chemicals, hypodermics, scalpels. He fished out his prey of the night before from jars filled with chlorobutanol, a mixture which kills but leaves the muscles relaxed, and went to work deftly with the injections of preservatives. The embalmed victims then went into another container duly labeled with tags stating date and location. The Indians leaned in a little closer when the scalpel was produced and the lizard dissected in a trice. Braulio was chattering incessantly, but his hands moved with speed and precision. For the Indians the performance was like that of the little girls in the curiara for us—a vision of expertise. The word was spread that day, because when night came, the packed benches were alive with nods and whispers and pointing fingers.

Whenever Braulio stood up or moved around, there were big smiles and a murmured "Torumo, torumo." Charles clapped him on the shoulder.

"You've been honored. They've given you a name and that's very rare with the Makiritare."

Braulio was pleased. Perhaps he was thinking of his astonishing feat of marksmanship that afternoon and wondering if the Indians had heard of and remembered William Tell when they had chosen his title.

"So I'm Torumo. I'm very glad. What's it mean?" he asked, smiling.

"Frog," said Charles.

But the Indians can be imaginative as well as snide, and the next day they were to give us more than a nickname to take away with us to Sarisariñama. In between journeys to and from the airstrip with loads for the shuttling helicopter, I noticed that a group of elders, one of whom spoke Spanish, were in serious conversation with Professor José María Cruxent, a lone archeologist who had dropped in on us on his way up river to search for traces of an older, more advanced culture than that of the Makiritare. The professor, a Catalan, was a sight for sore eyes with his jungle boots, sturdy bare sunburned legs, shorts, pipe and a battered solar topee of the type favored by the British sahib in old India. I was intrigued to find him taking notes and noticed that the elders had become sonorous, and their gestures biblical. Something was going on. I edged up to the group several times between relaying packs but they were too engrossed to notice—or had noticed only too well that I was being nosy. But when the word came from the pilots that Jimmy and I were to leave the fleshpots of Sta. María and fly up to the plateau to scout for water, I seized my chance. A polite farewell was in order, after all.

"*Con permiso,* professor. I would like to say good-bye."

He was about to stand up and I pushed him down again with deference but with determination too. I was not to be deterred. I pointed to his notebook:

"Something interesting, professor?"

He considered me a moment, a professional scientist weighing up a worm of a layman, and then relented.

"You understand, Señor Nott, that no Indian has ever been on top of Sarisariñama?"

"Of course."

"So they have never seen and cannot possibly have known of the existence of the holes?"

"Correct."

"Well listen to this. They say that it is a part of ancient Makiritare lore that once upon a time the hole was the lair of a monster named Cuyakiare, as tall as twenty men. It would come down from the summit in the dark, in the rains of the storms, and seize and eat the tribes. It lived many years in the hole with its father, Urorewe-vaka, until one day they perished in a fire, lit by themselves."

He gave me a long, steady look, warning me to stay solemn.

"Thank you for telling me. I'm going up on the next flight and hope to climb down into the hole tomorrow. I'm glad what's-his-name and his old man got cremated."

The professor looked at me without expression.

"The Indians say you can burn the dead, but you can't burn their *spirits*," he said.

4

DOWN IN THE DARK AT THE BOTTOM OF THE PIT AN eerie screech rang out above us, echoing round the walls, and then a wild clamor of them, strident and harsh. And in the din there was another sound. A hard click-clicking like the snapping jaws of some flying beast.

"Bloody pterodactyls!" I shouted.

We leaped to our feet, delving frantically into our packs for the flashlights, grabbing for our piton hammers. We beamed our lights up the wall and through the trees, and backed against the rock, ready to fight off the monsters. The wingbeats were thick and fierce now, and only yards from our heads. Instinctively we put an arm over our faces.

And then we caught them in the beams. They were three feet across, but with feathers, not leathery wings.

67

"Guácharos," Charles shouted. Soon the whole colony, maybe three or four dozen, was wheeling and diving around us, croaking and chink-chinking in outrage.

We lowered our hammers and switched off the flashlights, leaning back against the rock and breathing deeply to bring our pulses back to normal. Well, that's a dramatic start, I thought. Our first night in the time capsule and we go back to the Age of the Dinosaurs, to the screams and screeches of its giant, flying reptiles. And what screeches! Amplified by the huge sounding board of the rock face and then multiplied tenfold by the echoes from the walls that enclosed us.

"Pterodactyls!" said Charles in disgust.

He need not have been so disparaging of the guácharos. Not everyone gets so close to these birds. They live only in this region of South America and in Trinidad. They nest in caves or cracks in the rock and emerge at night to hunt for food. The clicking noise is their radar; the echo of the sound warns them when they are flying into obstacles. We could not know it at that moment but they were the fount of what was to be one of our strangest discoveries in The Pit.

Meantime they had scared the hell out of us and taken our minds off our situation. In unspoken accord we got moving with the bivouac. Jimmy dug out a carbide lamp and, with me reading out the damp, bleary instruction sheet he poured in the requisite number of carbide pellets, screwed the cap on and then topped it up with water. In the center of the reflector is a flint and milled-edge wheel, like that on a cigarette lighter. You strike the wheel and it lights or doesn't light, depending what sort of day you're having. Jimmy struck the wheel and with an evil hiss the gas enveloped us. There was no escape. One step away from the bivouac site in the dark and you'd break a leg. We squatted against the wall, gasping. It was our first reminder of industrial miasma since the fume-belching tractor in Cacurí, and here it seemed an especial affront to the primeval world we had penetrated. I peered at the instructions again.

"It says here, Jimmy, that if the user doesn't light the lamp on his second try there will be a hundred-ton rockfall on top of his head." We emptied the lamp and began again. This time Jimmy held his hand over the reflector to capture the gas, struck the flint-wheel, and got a neat two-inch flame.

"Well, we've had the *son.* Here's the *lumière,*" I said, hopefully.

Not a smile from anyone. They were staring in disgust at the feeble tongue of light. Just a moment, I said, with a gesture of the hand. I hammered a piton into a tree and hung up the lamp to face the rock and there was immediately a glow we could live by.

"A Somali soldier showed me that trick in the wilds of the Ogaden," I said loftily.

"What trick?"

"Well, you point your lamp at something with a pale surface, like the rock, and it suffuses."

"It what?"

"It su . . ." at this point I bent suspiciously to peer at the other two. They were at it again. Taking the mickey. Sarky pair of coves.

I stood up sniffily: "The way certain not-very-well-known comedians are performing tonight, an observer might be forgiven for not knowing that we are *entombed.*"

That shut them up. It shut me up too, as a matter of fact. We went on with our jobs in silence and soon had a reasonable bivouac arranged. Charles had his short jungle hammock slung between two trees, Jimmy between one of the trees and a piton driven into the rock. I laid my sleeping bag in a narrow gutter of dry leaves up against the wall. Charles then very sensibly insisted that we make a full meal, however dead-beat we were. We lit the small butane stove, contributed water from each of our canteens and came up with rice, nuts, Spam and dried fruit.

The guácharos had flown off by now and in the ancient stillness of The Pit, undisturbed by the faintest wind, we reviewed our situation. We have gone over the conversation of that first night many times, in our own unscientific way, trying to add our little bit to the record of how men behave when faced with the strong possibility of their own extinction. It was the same as I'd always found it; until extinction is an immediate prospect, there are no heroics, and no despair. At best it was a desultory debate. Conclusion: We're trapped. Resolution: let's leave it till tomorrow. Tomorrow, as it turned out, another button was to be pressed in our time capsule, which shot us forward from the Age of the Dinosaurs by millions of years.

We had made several reconnaissance flights over the hole in the previous months and taken hundreds of pictures. Studying these, we had picked the route we had descended the day before as the least difficult and dangerous of three that were possible. It was an upside-down process, because normally the climber approaches his objective and has time to study it from below before making an attempt at it. All we had had were brief glimpses of the walls from above as we shot over them in small planes. The pictures were of course the same; they gave us a view downward instead of a view upward. But it is an axiom among climbers that you cannot really tell whether a route will "go" until you have got close enough to it to rub your nose on it. I explained this to the others as we ate our horrible fortified milk powder for breakfast.

"While we are getting on with the job I'll be studying every yard of the walls. You never know. There may be a route we can climb."

Jimmy pointed to the wall in the southern arc of the hole, more than 700 feet high and all of it overhanging.

"If you think you're going to get me on anything like that, you're out of your head."

"No, nitwit. We'd all have to be Grade Six aces to try that and anyway we don't have the equipment. But if you look through the trees to the northeast arc, you'll see that it's only vertical. There may be something there. So eat your mash and shut up."

I noticed Charles had nothing to say and began to suspect he was cooking one of his famous, far-out but often eventually feasible ideas. He was. And later it was to cost us a month's hard labor concentrated into 24 hours.

But for the moment it was enough for our peculiar brand of tight-corner psychology that we would at least be looking at the problem during the day, and we said no more about it.

In any case, Charles was going over his checklist. Movie camera, still camera, specimen sacks, waterproof surveyor's notebook, caver's headlamps, and a 1,000-foot reel of nylon thread. This last item was to be our guideline for our return along the tortuous course of a supposed underground river we were going to try to follow. The river was one of those theoretical answers to one of the major questions we had to answer. Where does the thousands of tons of annual rainfall go in The Pit?

Does it filter through into an endless honeycomb lower down inside the plateau? Does it drain into underground galleries deep below the bottom of the hole? Does it run off through subterranean tunnels to connect with the smaller hole about a mile to the south? If this last was correct, the place to find the tunnel was in the cave under the guácharo cracks which we had seen in the aerial photographs.

It was 9:30 A.M. and the sun's rays were creeping down the western arc of the wall. It was an unconscionably late start but we were still stiff and tired from the exertions of the descent.

"Okay, *inglés*. You're the routefinder. Let's go . . . *vámonos*," said Charles.

I picked up a coiled rope and we skirted the foot of the wall for about 50 yards, then slithered down a series of gullies and short drops to the entrance of the cave about 200 feet below. Above us the huge wall sloped outwards the whole of its height, its top overhanging us by at least 100 feet. Below us in the bowl of The Pit, we could see dark, menacing crevasses between great jungle-covered boulders. It was our first glimpse of what awaited us when the time came to cross the bottom of The Pit to the other side. But the cave was an immediate disappointment. It was blind. No more than a recess, like a chapel in the side wall of a huge cathedral. It narrowed as it rose and finally closed to form the tall, vertical crack where the guácharos nested. I studied this crack with relish, calculating where the pitons would go, how to surmount each overhang . . . until Jimmy tipped my helmet over my eyes.

"Forget it. This isn't a climbers' club meet," he said.

From the floor of the cave a scree slope ran down about 120 feet to vanish into the crevasses between the rocks below. But when we crossed to it we found it was a scree slope like no other on earth. It was Wonder Number One of The Pit. We crouched in astonishment to examine it. It wasn't scree at all. What we had thought were the small stones that form such slopes were not mineral, they were vegetable. They were nuts and seeds and fruit-pits. Millions upon millions of them. Not a pebble among them. We christened it immediately: Nut Mountain. We picked out five types, some from the Euterpe palm, some of the Jessenia genera but of unknown species, some we couldn't identify. The biggest were about an inch and a half long, with a fibrous skin. The smallest were barbed yellow seeds like a common burr. We sat down to think it out. Charles explained that the guácharos eat the fruit during their nocturnal foraging and then, once back in their nests, regurgitate the stones or seeds. The slope ran down about 120 feet at an angle of about 40 degrees. It was 30 feet wide on average. But how many seeds are there in this unearthly pile? And if the colony of birds has been more or less constant in numbers, how many thousands of years has it taken them to build it? Another question: there were feathers and fragments of eggshells in among the seeds but no droppings. The rock, instead of being plastered, was clean. Who housebroke these guácharos, Jimmy demanded to know. We still don't know.

Charles, already eyeing the walls of the cave for geological specimens, handed me a caver's headlamp from his pack.

"Why don't you try to find the tunnel while we work up here?" he said.

I clipped the lamp onto my helmet and started to trundle down the slope. In seconds I sank to my chest in the nuts and was careering down in a hissing avalanche of them to shoot into the first crevasse at the bottom. With the Brewers cracking at the rock above with their hammers, I squirmed down a twisting shaft to listen for the sound of moving water. Nothing. I climbed up again, crossed to another gaping black hole and squeezed down it as far as I could go. Not a sound. An hour later I was still at it, wriggling down the cracks, crawling horizontally a few yards to a dead end, dropping pebbles down fissures too narrow to get into, and listening to their tinkling fall. None of them went very far, and there still was no sound of water. Not one drip. So much for the legendary river, I thought, lying flat in the dark in an 18-inch-high tunnel. There's no water at all. And that brought an altogether different kind of speculation.

I crawled somberly out of those intricate grottoes, losing my way once or twice, and emerged into the light at the bottom of the slope. The Brewer Bros. Ltd., were still blissfully engrossed in their rock collection as I stumbled and swam my way up Nut Mountain. They turned to me eagerly.

"Did you find it? The tunnel? And the underground river?"

"No tunnel. No river."

They digested this in silent disappointment. We had dearly longed to penetrate that black river, full of 30-foot eyeless newts, weird amphibians, and giant bats. If we caught one, we'd be on the talk shows.

"There's no water either," I said. There was no response. They are spoiling the essence of this dramatic moment, I thought. I tried another tack.

"Look at it this way. How much water have you both got?"

"About a third of a bottle."

"Well?"

"Well?" They replied, still not with it.

"If there's no water, how long can we survive?"

Our aerial photographs had shown thin waterfalls tumbling into The Pit at several points. We were aware that they had been taken in the rainy season and that we would climb into it in February, in the height of the dry season. I had argued that we should take at least a five-gallon tank down with us but had been outvoted. Indeed it was difficult to imagine that a hole of this size, with more than ten acres of jungle at the bottom, could be dry. I had dire memories of this sort of problem. Three Americans and I had made the first ascent of the 3,000 foot wall of the Angel Falls in January 1971, in the dry season, and had hauled ten gallons of water with us. But of the ten days it took us to climb the wall it rained fiercely for eight and a half. With these plateaus, you can't win.

It was now about noon and we had to fix our priorities. Charles insisted that he and Jimmy finish their geological survey because we would not be returning to the cave. This involved hammering off flakes of rock, numbering both the flake and the hole it came from and photographing the whole process so that later in their laboratories, the geologists could examine the flakes and study them against the pictures of the rock. They had a lot more work to do on the several strata that could be reached.

We arranged that I would leave them there and go off to try to complete three tasks. First, to cut and equip a clean, more direct route up to the bivouac so they could get their sacks of specimens up there. Second, to prospect beyond the bivouac for water. Third, to attempt to find a

route straight down from the bivouac northeast towards
the center of The Pit, again, searching for water on the
way. This was a tall order for one afternoon so I grabbed
the machete and with more fervor than science, tore into
the jungly walls blocking the way to the bivouac. Now
it's one thing to be a machetero—machete man—when
the rest of the party are following along behind, and
another when you have to leave a trail that will be
unmistakably visible for a party which will follow on
later. My method is, as Jimmy said later, one of total
destruction. Cut everything from eye level down to the
ground. This last is the nub of the trick. Where there are
leaves on runner plants underfoot, slash them off with
sideswipes of your machete. The pale underside of the
upturned leaves point the way even in the gloomiest
forest. My upward trail was a masterpiece. Every loose
rock was heaved off down the slope, roots were dug out
for handholds in the gullies, and every last living leaf and
stem hacked off. The trouble was, my companions
complained, there was nothing left to clutch at to keep in
balance on the awkward up-and-down steps. Sweating
and proud, I reached the bivouac and sat down to survey
the mess of scattered pots, rations, sleeping bags and
whatever. What a place to die of thirst in, I thought. At
that moment I became conscious of a sound that was
familiar but not quite focused in my memory. Tap-tap. It
sounded like someone flicking his finger on a newspaper.
The London *Daily Telegraph*, no doubt. I stood up,
turning this way and that to find it's direction. It was
along the wall beyond the bivouac. There was no way
straight to it because there was a deep shaft a yard away
from our site. I dropped down a short wall into the
undergrowth below the bivouac and circled round. After
50 feet I came upon a clearing between the trees and the

wall where there was a tree, fallen from the rim of the hole, smashed into splinters on a pile of boulders, part of the daily bombardment The Pit was treating us to. But also falling onto the boulders was a shower of water drops from an overhang about 150 feet up. I looked up at this projection and saw it was mossy and slimy.

There must be a spring there, I thought exultantly. I scrambled back to the bivouac and dug out the three six-foot-square plastic sheets I had brought as rain-catchers. Back next to the smashed tree I glanced up suspiciously to see if there were any others following it, and carefully arranged the sheets as basins among the boulders. By the time I'd got the third one fixed the first had half a pint in it. It was a brown liquid and as full of floating objects as a baby aquarium. But it was sweet. It was all the water we needed.

I looked up in time to dodge a couple of projectiles from the surface, a clod of earth and a stone which shattered noisily on the rocks, and scrambled cockily back to the bivouac. Recalling my trail, and now the water supply, I reckoned we were getting the measure of The Pit. Except, of course, that it had us trapped. With such luck, I'd surely find an autobahn leading to the center of the bowl, my third job. But in a whole hour of probing through the jungle, I found no way which was not cut off by steep cliffs dropping down to the lower levels of the basin. I could put fixed ropes down them but that would mean we would have to jumar up each of them and then haul up our sacks of specimens. Time-consuming and very hard work.

It was then that I heard another new noise. A faraway murmur I couldn't place. In The Pit it was difficult to tell where any noise was coming from, even whether it was above or below you. I shrugged it off and decided the only way into the center of the bowl was to go halfway down the trail to the cave and then cut left. I resolved to make a start on this and climbed back up to the bivouac. I could still hear the murmur and it was becoming more distinct. I was nonplussed. In fact, a time-capsule button had been pressed and we were about to be flipped forward over the eons into the present.

I crossed through the bivouac, followed the 50-yard ledge, and started down the first gully of the cave-route, admiring my handiwork on the way. I heard voices and glanced down to see Charles and Jimmy climbing up toward me. We met up on a flat, pink sandstone ledge, all bursting with news and expedition gossip. And then all at once we saw it, and jumped a foot in stark terror.

A few feet above our heads a bulbous, muddy green object floated. In total silence it moved infinitely slowly to one side and I saw that it was topped by what looked like a conical metal cage, like a buoy lifted out of the sea. The Thing paused and moved back above us. The murmur had stopped. I blinked and shook my head, but that macabre UFO was still there.

"It's the winch!" Charles shouted. "They've got it rigged! That's the test load. Look at the rope!"

And there it was indeed, a green nylon rope tied to the cone and stretching upwards more than 700 feet to the overhanging rim of the wall. The object was a duffel bag. It was suspended, not floating, and the sideways movement was its slow, gentle swing.

How to get at it? On our left a squat rock tower struck up from the slope. We scrambled up it, tied a knot in the end of our climbing rope and threw it in an arc round the cone. The weight of the knot brought the rope round to us and we caught it and pulled the bag towards us until we could grab it. We unhitched the load and anchored the cone to the rock tower so it could not swing away out of reach. Charles reached into his pack and pulled out our walkie-talkie. He extended the aerial and pressed the *listen* button on the microphone. Immediately the stillness was broken by a remote voice.

". . . receiving me? Can you hear me? Shift the button to the speak position and answer. Over."

Charles put on his radio voice (we've all got one) and replied. "*Carajo, hombre!* You're loud and clear. *Felicitaciones* . . . congratulations! I *told* the bloody *inglés* you could do it! Over."

The voice crackled back: "Tell the bloody *inglés* to blank blank his blank."

Used to small, bare-bones expeditions, I had scoffed a great deal during the months of preparation about the plans for communication and the winch. I never believed they would get that machine rigged on the edge of The Pit and still less get it to run free up the overhangs. Now the Brewers were staring fixedly at me, one eyebrow raised, waiting in triumph for me to repent. I didn't, then, although my ears may have reddened a bit. But I was glad to do so two days later. Charles returned to the microphone and explained they'd have to run out more rope so the loads would touch down where we could reach them.

He moved the button to the *listen* position and Lionel's voice came again: "Roger. Now hitch the empty bag to the cone and stand by. We're going to take it up and lower you a cable. Out."

A cable? The grinding murmur started again and we

watched a full ten minutes as the bag slowly rose to the rim, twirling incessantly on the end of its rope. After a moment a heavy branch whistled down to smash to matchwood on the ledge. Winch Handling Lesson No. One. Don't stand under the line of descent when loads are moving.

We climbed up into the gully and sat in silence until the radio crackled again.

"We are lowering the cable. There's a stone tied on the end to weight it. When it reaches you, take it as far to one side as you can so it won't interfere with the next winch load to come down. Over."

"Understood. Out," said Charles.

Understood? A cable? With a stone on the end? What in the hell is going on? I was bursting with questions but one glance at Charles's expression held me off. He was trying out his deadpan look but there was the faintest trace of the supercilious about him and I wasn't going to give him the chance for some more one-upmanship. Jimmy was easier to read. He was staring off into space grinning gleefully at nothing and no one. I waited. This was a plot. Eventually a three-inch stone lowered itself into our view, tied on the end of a minutely thin cable, hardly more than a thread. Jimmy grabbed it and showed it to me.

"There it is. The cable. D'you see? Feel it!"

"Um," I said.

Charles took the radio.

"Hello, Lionel? We've got the cable. We'll take it to one side and call you back when it's clear."

"Roger. Out."

Jimmy tied the end of the cable to his waist loop and started up the gully, flipping it over trees behind him, or, where the trees were too high, pulling it carefully through the foliage. As soon as he reached the level stretch along the foot of the wall he shouted down to us and Charles flipped the radio switch to tell the surface team it was clear.

In minutes the murmur began and soon we could pick out the load coming slowly down through the overhangs. When it arrived we unhitched it and replaced it with a load of rock specimens. Charles radioed the surface party to haul it in and said we would return to the bivouac and call back in half an hour. He climbed slowly up the route with the radio and I followed him with the winch load, hefting it suspiciously in my hand but saying nothing.

At the bivouac Jimmy had tied the cable to a tree, with its free end hanging to the ground. There was a certain feeling in the air, a sense of delicious anticipation which I didn't like one bit. What's more, I was peeved that with all these goings on there hadn't been an appropriate moment for me to announce, nonchalantly, that with my superb jungle skills I'd found water and saved the expedition from thirsty death. After a moment's plotting I brightened up. I'd upstage these technological twiddlers yet. I stealthily extracted the collapsible one-gallon water tank from a pack, and slid off in the gathering gloom to my as yet unannounced reservoir. Glancing fearfully up the darkening wall for projectiles, I filled the tank from the plastic sheets and stole back to the bivouac. The others had the lamp lit and were bent over something or other at the foot of the tree. I swiftly lit the butane stove, put a pot of water on and dug out the tea bags and the sugar. I'll show 'em, I thought. When the water boiled I flipped a tea bag into a bowl, poured in the water and held it out. My big moment had come.

"Tea, anyone?" I trilled.

But at that moment something else trilled, and I fell back against the wall aghast, my tea forgotten. The buggers were twirling a little handle and the trilling tinkle was coming from the object attached to the cable.

It was a *telephone*.

I stared in apology around the dark Pit and then upwards to the circle of stars encompassed by its rim. Hundreds of millions of years of peace were written there. And there was Charlie, twirling his handle and barking into the mouthpiece.

"Pit to Camp Two. Are you receiving me. Over."

It sounded like the forward observation post of a howitzer regiment. But they hadn't finished with my tender sensibilities yet. That night I was really going to get trampled on.

Charles again: "Put me through to Archie in Camp One." Through to Archie? Across a mile of jungle? But he got him, checked how the rations were going, the water, the fuel, the helicopter and so on, and then talked one by one with the scientists. So all right, I shrugged, they have rigged a radio link between Camp Two, which was where the winch had been set up, and Camp One, and somehow connected it with our field telephone.

"Okay. Get the Centro to connect with 357996." Charles was looking at me steadily. And then suddenly he was shouting. "Hello, hello! María-Mercedes? How are you? . . . And the children?"

I swear to you, Charles was talking to his wife in Caracas. The Centro was the office of Fundasocial, a civilian rescue and emergency organization in the capital. Lionel and Federico were linked to them by Stoner single side-band radio on frequency 6730. They in turn could connect our system with numbers in the city. So there we were trapped in The Pit, with all its perils, and

Charles was asking his missus about the kids. Fascinated, I stored it in my mind to add to the men-in-danger records. I thought of the crew of a sunken submarine tapping jaunty messages to rescue divers outside, separated from them only by the thickness of the hull, but as far from safety as ourselves, blocked from escape by a rotting cornice in which our rope was buried. The dank still air of The Pit must have been conducive to telepathy, for Charles was thinking about escape too. He replaced the telephone in its box and harrumphed for attention. He was about to announce a very hairy scheme: Operation Exit.

"You want to get out of here?" he said.

"Yes."

"Well, we're going to fly out."

Jimmy made flapping motions with his arms, practicing. But Charles went on unperturbed and with total conviction.

"Tomorrow we go to the center of the hole. We cut down a circle of the tallest trees. The helicopter will spiral down into the hole and drop the ladder to us. We climb up and out we go."

We studied him in silence a moment, resisting the urge to take his pulse.

"Spiral down into the hole? That $200,000 helicopter? Skimming round these walls?" I said.

"Well, he can come in at an angle, if he wants."

"And how do we cut down those trees?" asked Jimmy.

"We'll get the winch to lower us the chain saw."

"Chain saw? We're going to lug a chain saw into the center of the hole? Across those crevasses?"

"Yes."

"Suppose the trees in the center are growing so thick together they won't fall when we cut them?"

"Then we'll cut them in sections."

"Do you realize you can't jump out of the way of a falling tree in there? If you sidestep, you're likely to fall down a shaft and break a leg?"

"Do you want to get out of here or don't you?" Charles said huffily.

The argument rolled on as we boiled up a pan of lentils for supper and followed it with my abandoned tea. Charles was unshakably confident that he could persuade the pilot to make the attempt. But for the life of me I couldn't see the helicopter coming down into The Pit.

The telephone tinkled and Jimmy picked it up, listened and handed it to me:

"Call from London," he said, nonchalantly.

"From London," I echoed, vapidly.

I took the receiver. Lionel was on the other end.

"A news agency in London is talking to the Centro but we can't connect them with you. What shall I tell them to say?"

I thought of my dear old mum in England, always worried about her forty-five-year-old lad, reading some horrible headline in the morning newspaper. One had to be careful with one's mum.

"Anyone in the Centro speak English?" I asked.

"Yes."

"Tell them to say our escape is *problematic.*"

I should have known better. My cautious phrasing was enough for the news hawks. The next day there were some choice banner headlines in the world's press:

THREE EXPLORERS TRAPPED
IN LOST WORLD

EXPLORERS SWALLOWED BY
ISLAND OF TIME

Poor Mum. And poor us. Tomorrow was Slave Labor Day.

5

Dawn in the pit was like the first dawn on earth, pure, still, limpid. I had an uncanny sense that last night could have been a thousand million years long for we woke up in a world that had not changed since its beginning. This tangled jungle was here before the Himalayas and the Rockies had thrust up from the planet's surface; these silent insect legions were creeping here before the sands inched over the green of the ancient Sahara. The same pale blue circle of sky had looked down into The Pit millions of years before the great reptiles came and millions of years after their extinction. It must have blinked a bit that morning, for here was a change like none it had looked down on before. Three scruffy homunculi had appeared in a flash of time. They were talking of heliports, radio communications, angles of approach, and escape. This last word seemed to mean a lot to them, which was strange, for who would wish to leave the supreme peace of this place?

"I want to get out of here," Jimmy was yelling over his breakfast.

"It's creepy. It's not peaceful at all, it's tensed up. The whole place is just waiting for something that's going to happen."

"Something involving us?"

"What do you think? We are foreign bodies in here. Like bacteria. The Pit is going to react. One of the damn walls will crumble and bury us. Or the whole bottom of the bowl will drop through into some huge cavern beneath. There'll be a poisonous plant . . ."

"More cheese Jimmy? To shut you up?"

But he was right to try to get us moving for we had a great deal to do. We sat messing about fitfully with our breakfast rations. I was taking my time because I was smoking my last cigarette, a fateful moment for an addict so far from the shops. But the others were not exactly straining at the leash either. I prodded Charles with my toe.

"What's wrong with you, ball of fire? You haven't even got your boots on yet."

He stretched himself gloomily.

"I don't feel any livelier than I did yesterday morning. But we had an excuse then. We were stiff from the descent."

We looked at each other in startled discovery. We were all feeling an unaccountable lethargy. Jimmy proposed that it was not physical but psychological, that the insidious tentacles of The Pit were feeling their way into our brains, draining us of fight, sapping our will to survive.

"They'll find us here as three skeletons, stretched out where we lay, too tired to try to escape."

We decided in the end that the air seemed pure because it was cool, but that in reality it was dank and heavy. We were told much later that there may have been a concentration of carbonic gas down there which would account for our enervation. I'm not sure how much science there is in the suggestion, but it sounds pleasingly dramatic and I wish we'd known it at the time—Jimmy and I could really have put on a pantomime.

Anyway, thoroughly gassed, we staggered finally to our feet and made our way along the flat stretch, then down to the winch loading point. We got out from under this very rapidly and headed left down into the unknown territory of the bowl. The sequence was the same; I went ahead with the machete and the other two followed, carefully photographing and filming specimens in place, then collecting them in the sacks.

Charles, adamantly thorough as always about this work, paused here and there to make sketches and enter his observations in a notebook. It is a telling illustration of the type of terrain we were in that we were never separated by more than a few yards. No matter how slow the others were, I never got ahead because the route was so difficult and tortuous my speed slowed at times to a yard a minute. The floor of the bowl was formed by the rocks that had fallen there when the surface of the plateau caved in in some remote age of history. The boulders ranged from the size of a pillow to monsters thousands of tons in weight. Between them there were crevasses and twisting shafts sinking down into darkness. At times the route dipped down these shafts and followed a tunnel to emerge again vertically upwards the other side of a mansion-sized block. Every inch of these rocks was slimy and treacherous and over every inch the relentless green jungle grew, invading every cranny and crevice, covering the narrower shafts with a traitorous mat of roots and leaves, and at eye level so thick it was impossible to prospect one's way ahead more than three or four feet at a time. Indeed, to keep on the general line toward the center of the bowl I had to keep glancing upwards for a glimpse of the walls around us through the foliage. Up there, there was always an arc of the circle of rock glowing with sunlight, and how we ached to feel those rays on our faces down there in the green gloom!

But I'll admit to feeling a certain amount of swagger to be up against this lot, with my machete swinging. I was absorbed and happy probing this way and that, trying to thread a reasonable line through the maze that would not require ropes and pitons, a line that either myself or the others could follow unhesitatingly back to the bivouac, and then retrace the next day or the next or whenever we had to come back that way. Many of the rocks were covered with a pretty, white bell-shaped flower growing on a runner plant. I must have swept scores of them off their stems in order to mark the route. I was reluctant about this mayhem at the time, and you can imagine my feelings months later when I was told by Julian that the flowers I was destroying were new to science.

I think we were all happy at this time, lethargic or not. None of us thought of our real situation. The fascination of being on new ground, the knowledge that so much of what was being collected was unknown, had driven our basic purpose from our minds, until after two hours to cover about 130 yards, Charles said:

"That's it. There's our heliport."

The Pit at an angle from the air. Jaua plateau
in the distance.

Even before the descent there were tricky moments.
Nott entangled in a tree near the edge of the hole,
after dropping by rope from the helicopter seen
hovering above.

Nott and James Brewer
preparing rope and
rope-ladder for the drop
out of helicopter.

Left James Brewer rappels
down rope.

Looking up from the
bottom of the 900-foot
deep hole.

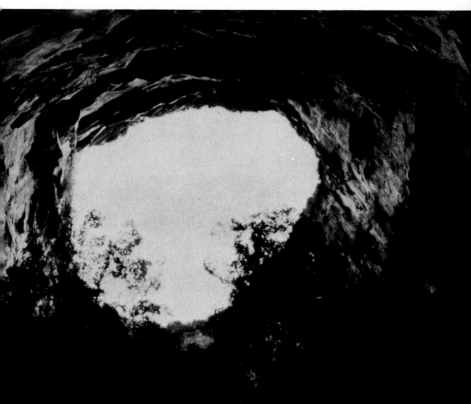

Charles Brewer in right-hand corner, looking lost
in The Lost World.

Indians of the Makiritare tribe, in Santa Maria de Erebato, with communal house in background.

Makiritares burning out a tree for a dugout, or curiara, at Santa Maria de Erebato, an expedition staging post.

A load of specimens goes up the 800-foot winch cable near the bivouac site.

Nott fells a tree with a buzz-saw in the center of the hole, an abortive attempt to persuade the helicopter to spiral down into the hole and drop a rope for our escape. The pilot decided it was suicidal.

James Brewer, *left,* Nott with machete, and Charles Brewer waiting for water drops from the rocks above to collect in the plastic sheet right. This was our water supply down the hole. The splintered tree fell from the surface the day before, part of a daily bombardment. (Timed photo taken with camera perched on a rock.)

This dragon-fly, which may be an unknown species, was found in The Pit and on the Jaua.

A carnivorous plant folds its sticky tentacles round a big jungle ant which will be slowly digested.
The plant is of the *drosera* species and grows in abundance down in The Pit.

Left to right: Geologist Antonio Quesada, zoologist Braulio Orejas and David Nott, at Camp One on top of Sarisariñama.

Charles Brewer took this picture while perched on the electron wire ladder by which we got up the last 200 feet of the wall.

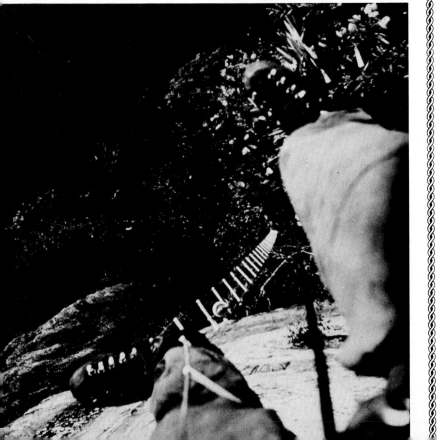

I was cutting round the base of a huge house-sized conical rock and looked back at him in wonder, machete poised. He was pointing upward.

"We have got to get on top of this thing and clear it. That's the solution. We won't have to cut down so many trees."

I retraced my steps a couple of yards and scrambled 20 feet up to the top of the rock. It was fairly flat up there with about half a dozen big trees growing, and all as impossibly tangled as it was below. I looked up around the walls and we seemed to be near the center of the bowl. Charles and Jimmy came up and we studied the position. Our rock—it was merely the topmost tip of some colossal monolith—would need some tricky lumber-jacking to clear. Then we could count a good dozen 80-foot trees surrounding the rock that would have to come down too. Each one presented a problem of approach. Just getting to their base would need some hard machete work, and some were surrounded by a plant so dense it could not be cut through. We would have to force our way through it on hands and knees. But this was only part of the hazard. We could see at a glance that the major danger was the lack of any solid, level foothold in this jumble. I had a vivid picture of us unstably perched on slimy rocks, crevices all around us, up to the knees in roots and lianas, and manhandling a demon buzz-saw. One slip and we would amputate something. One mistake in a tree's line of fall and we'd be too tangled to get out of the way, or would make a leap for it, land in a crevasse and snap a leg. There was, of course, no hope at all for an injured man down in The Pit. He would really be down there for good.

But Charles was actually rubbing his hands with glee, or something. He took the walkie-talkie out of his pack and called the surface. "Hello, Camp Two. We've found a heliport! We need the saw, start it on it's way down now and we'll start back to get it. Out."

Charles was shaking his head, whistling softly to himself.

"Just wait till you hear that thing down here," he said.

With the trail cut, we got back to the point under the winch in 20 minutes. The load had already touched down and looked menacingly big and lumpy to me. We stripped off the sacking and surveyed our tools in dismay. There was the saw, 30 pounds of wickedness, a five-gallon plastic tank of gas—square-shaped, and a half-gallon can of oil with two holes already punctured in the top. Even Cuyakiare himself couldn't have wished a worse set of loads for that trail. Charles was encumbered enough with the movie camera so the three objects were left to Jimmy and myself. I looped a length of nylon thread to the handle of the gas tank and used it as a shoulder strap. Then I picked up that preposterous, leaking can of oil and the machete, and led off with Jimmy wrestling the saw along behind me.

That was not the merriest of trips. We changed loads every ten minutes to avoid being driven stark bonkers. We got into several maneuvers that would have put Laurel and Hardy out of business, but it was all deadly serious to us. At one point we had to ease down a thin, sloping tree trunk bridging a crevasse of unknown depth. A tree still with all its myriad, prickly branches on it, springy and malicious. We edged along its length, passing the saw along, the gasoline and the accursed, dribbling can of oil. At what we called the Leap in the

Dark, which was an underground passage severed by a crevasse which needed a five-foot jump to get across, life became really trying. You can't jump five feet with an open can of oil; you can't throw this offending object to your mate across the gap. What do you do? You climb down the crevasse, straddle your legs across it with one foot on each wall, then, with your mate holding you by the rope, you carefully transfer the object from one side to the other, then climb out the other side to join it. Our lethargy was growing chronic by now, and we dreamed of nothing closer to heaven than to simply collapse where we stood, and fade out.

Of course, the moment had to come when the oil can broke loose. Jimmy dropped it down a slimy 20-foot deep hole. I looked at him commiseratingly. Poor old chap has to climb down there and get it, I thought. But Jimmy didn't move. He was staring steadily at me with the oddest expression. I studied it slowly, interpreting its varied meanings. I was astonished.

It read:

"I dropped the can and I should climb down and get it. We are both gassed into torpor but you can get down that horrible hole better than I can, so *you* go and get it. I'm too inert to give a damn."

This was most unlike him. Jimmy was fit and, a good expedition man, always did his share of the work. He must be feeling pretty low, I thought, handing him the saw. It took me some time to get down the hole and much longer to climb out again with that weeping can. So long in fact that Charles appeared as I emerged and concluded I had been down there in our endless search for a way into the underground river.

"Did it lead anywhere? Did you find any water?" he asked eagerly.

"No, but there's oil down there."

Charles was dumb struck.

"Oil . . . ?" he whispered.

I held out my sticky hand to him.

"Yes. Feel it."

"My God, p-p-petroleum?"

"That's right," I said, producing the can. "Good stuff too. Just right for two-stroke engines."

Somewhat enlivened by this scene Jimmy and I pushed on through the maze until finally we reached the heliport-to-be. Our crash course in lumberjacking was about to begin.

I hefted the saw dubiously. "Have you ever used one of these, Jimmy?" I asked.

"No."

"Well, how do we cut a tree so that it falls away from the heliport—and away from us?"

Charles came up behind us. "You cut a notch like a wedge that goes halfway into the tree on the side you want it to fall. Then you cut through from the other side towards the point of the notch. There will be a loud crack and you run like hell."

"Run? The world record for 100 yards down here is 120 minutes."

"So you will just have to be dead right when you cut that notch, won't you?" he said comfortingly. "As for me, I'm going on ahead to where I can see what looks to be a very strange-looking orchid."

I held the saw while Jimmy yanked the cord to start the engine. It was the most unneighborly thing I ever heard and we looked at each other in alarm as the noise swelled up and echoed round the walls. I pressed the trigger on the saw's pistol-grip and jumped in fright as the chain-blade jerked into a blur of motion. I turned to the nearest tree, a 30-footer about six inches thick, and cautiously touched the saw to the bark. That damn machine went right through the trunk with one hideous shriek and the tree crashed down between us.

"We'll have to do better than that, won't we, mate?" I said, handing the saw to Jimmy.

Jimmy swiftly cut down two saplings to get the feel of it and then knocked off a bigger one. Heartened, we climbed up the rock and attacked the trees on top. Some, linked to their neighbors with a net of vines and lianas, refused to fall. We had to cut the neighbors too until the whole lot crashed down the sides of the rock cone. What with the screeching saw, the ripping lianas, the explosive cracks of the wood, and the dull *woomph* of the falls, the din in The Pit was horrendous. Jimmy, a sufferer from rockfall phobia, kept glancing fearfully up round the walls. I had told him that in high mountains avalanches and icefalls can sometimes be triggered by no more than a shout between climbers, and that we had better keep quiet if we didn't want to bring a thousand tons of rock down from the rim of The Pit. I didn't feel quite so glib about it now, and began to look around uncomfortably myself.

It was then I noticed what awaited us around the heliport. The surrounding trees were huge. They were unnatural too, distorted by their environment. All jungle trees must struggle upward for the sun, fighting for their place among those that surround them. But here in The Pit, where, depending on the season, the sun touched the bottom of the bowl only briefly each day, this struggle was desperate. The trees around us were far higher than they should be for their girth; they were a different race, grotesquely tall. It looked like the Wicked Wood in a Disney movie. One of them was only a yard from the rock, growing out of its own hole among the boulders.

I turned to Jimmy. "Sooner or later we've got to try to cut one of those big fellows down. And I can count twelve that will have to go. Why don't we have a crack at this one now and see whether we can manage them or not?"

We climbed down and examined our adversary. It was 18 inches thick and we had to lean out over the hole it was growing in to get at it. We took it in turns cutting the notch and, whether because of carbonic gas or the awkward position we didn't cut enough. I began to cut from the other side and got about 9 inches in when the saw stopped and failed to respond to the trigger. I had cut in almost horizontally and the weight of the 80-foot-tall trunk had settled on the blade. It was hopelessly jammed. I stared at it in confusion. Was this the end of the heliport? The end of our dream of flying out of The Pit like millionaire commuters?

Jimmy, more like his old self, pulled himself together and seized the machete. "Wedges," he said, like some magic incantation. He cut down a two-inch sapling with one swipe and in minutes had fashioned four 6-inch, sharp-ended sections. I found a handy stone about a foot long and we manfully labored a good half hour trying to drive our pre-Stone Age tools into the slit in the trunk with our pre-Stone Age hammer. It was hopeless. Our wedges splintered so badly we could have sold them for shaving brushes. There was only one way out of this, we decided. Cut the tree down with the machete. There was no room to swing this weapon in the narrow space between the tree and the rock and, on its other side we had to cut backhanded into the notch. The wood was red and hard.

The humidity was creeping toward 100 percent. We didn't just sweat, we streamed sweat. After an hour there was a loud crack and we braced to dodge the fall. But the tree didn't move. We cut again and soon there was another crack and still no move. I could have cut my nerve ends off with shears. Gritting our teeth we chopped on and then with a great bang the trunk snapped, the jagged end shot forward, grazing our ribs, and the monster fell. Right across the heliport.

We were getting quite cross with this damn tree by now, so we quickly retrieved the saw from down the hole where it had fallen, squeezed the trigger to find it was still its old vicious self, and climbed up the rock with it. We cut a 10-foot section off the thick end of the tree and let it crash down into the undergrowth below. Then we sawed through the trunk on the other side of the rock so that the long top-section fell too. The 20 feet that remained on top of the heliport we wrestled over the side like a broken spar from a galleon and it too plunged down to the foot of the rock. We were going great guns now, psyched up by the slaughter and dismemberment of the first of the giants, and we waded into the rest of the trees on top of the rock, handling the saw with vim and snap. We slashed and cut and chopped without pause and booted everything over the edge. When we finally stopped to survey the ruin we heard a muffled cry from Charles which we decided was coming from near the rock and somewhere towards the northeast. We scrambled over to that side but could see nothing. Nothing, that is, except a sea of flattened foliage and wrecked timber all around us, yards deep and impenetrable. Unwittingly, we had surrounded our heliport with defenses which made it the most difficult spot to get at in the whole Pit.

Down in the tangle we caught a glimpse of Charles's orange overalls as he crawled on his belly towards the rock. He emerged finally, scratched and bloody, panting with exertion, and climbed up to join us.

"What have you done?" he raved. "How the hell will we ever get our gear through this lot when the helicopter comes? And you've blocked off the whole route across the bowl. Give a pair of monkeys a saw and. . . ."

He broke off as a remote Tarzan cry echoed faintly round the rim of the hole above us. We listened as it was repeated. The winch team must be bellowing their lungs out in unison, we decided. Charles took the walkie-talkie out of his pack, pulled out the aerial and switched on.

"Hole to Camp Two. What's all the noise about? Over."

". . . can just see you. Wave your arms if you are listening to this message. Then transmit back."

We waved our arms and Charles repeated his message.

The voice came through again. "You are not transmitting. Try again and push the button hard. Wave if you have understood."

We waved and tried again.

Charles shifted the button to the listening position and the voice said, "There's something wrong with your set. Send it up before dark tonight. Now here's the message. Wave if you have understood. Our own radio is not working and we are out of contact with the helicopter and Camp One. We hope to repair it soon. The last message said they were short of fuel and were going to

Sta. María to pick up a barrel. Now for the questions. Do you need water?"

We stood still.

"Okay. We can see that you have cleared that rock nicely. But from up here it's obvious that you have to cut down a hell of a lot of trees round it. The helicopter will circle over to have a look when it gets back so you better get moving with that saw."

We waved our arms and without a word climbed down from the rock, forced a tunnel through the tangle and cut our way over to a group of four big trees south of the rock. We were working like robots now, hardly speaking, lugging the saw, the gas tank and the oil from tree to tree across the crevasses and holes and crawling to the rock every so often to get an elevated view of how we were doing. Slowly we moved round the rock and now had the big trees crashing outwards from the heliport like the spokes of a wheel. Many of them failed to crash through to the ground, and we had to cut our way behind them and saw down the trees that were holding them up. This was maddening work and dangerous too. We called it the Anti-Domino Theory. When one domino falls around here, the rest do not follow suit; they hold the bloody thing up.

This was like Devil's Island in the bad old days. Punitive labor in the dripping jungle.

Jimmy and I were crawling through the tunnel back to the rock when we heard the chopper faintly, and then its din blared suddenly down into The Pit as it crossed the rim and circled above us. Charles came wriggling through the tunnel like an adder and we scrambled up the rock and stood there waving our arms looking from the helicopter to the ring of destruction around us and back again to the gay white and red of our flying rescue machine. The pilot made another slow turn and we imagined him whistling in astonishment to his copilot at what three determined men can accomplish in five hours of relentless labor. He flew off south and in the silence Charles took out his radio. Sure enough, the surface team had anticipated us, and a voice came crackling through:

"Hold it. We are getting through to the helicopter now. Just a minute. . . ." The voice faltered a little, and coughed. "I'd like to make sure you're receiving me. Wave your arms if I'm clear."

We waved our arms.

"Uh . . . the pilot says you are in the wrong place. There's another protruding rock about 50 yards to your north. He says will you move there and—uh—start again."

The new rock was a tooth—a fang, if you like. It took us half an hour to reach and, The Pit jungle being what it is, we didn't see it until we bumped into it. The way up was on the far side and it was masked in flubbery, heavy, three-foot leaves with thick, sappy stems. My machete was dripping wet. It was like lopping off boneless arms and elephants' ears. Thunk. I got to the tip of the tooth and cleared it totally. Every leaf and twig. Plain bad temper, of course, and a yen for mayhem. But one tree held out. It was the tropical counterpart of a gorse bush, with a trunk as thick as my leg, quite the wrong vegetable to attack with a blunt machete.

I heard the others coming up the fang behind me and called to Jimmy, "Hand me the saw, will you. This thing is as tough as a steel hawser."

"The saw is down below and it's staying there until we eat."

So we ate. Dried fruit, nuts, hardtack biscuits and water. Perched up on the tip of the tooth we surveyed our new heliport. The first factor in the equation was that on the west side the tooth dropped down into a huge crevasse, 15 feet across and of unknown depth. If you stumble while trying to grab the rope ladder from the helicopter, don't fall that way, daddy, throw yourself down the other side where the drop is only 25 feet. All around us there were Big Ones we would have to saw down. To the south was a menacing clump of six rearing out of the tangle, their branches mingling, and swathed in a mantle of lianas and creepers.

"That bunch is going to be nasty," I said. "What's more, there are so many around this spot that we'll need another day of work." For some reason or other the nuts and dried prunes seemed to have gone straight to Charles' head.

"*Qué va, chico* . . . what rubbish! You see that bunch of four over there? I'll bet you I'll have them down in twenty minutes."

Jimmy and I stared at him in outrage. He hadn't even touched the saw all day. We studied the four Big Ones and then looked pityingly at our ebullient comrade.

"Right, mate. You lose and you wash the pots and get the water for the rest of our lives down here. You win and we'll feed you in your hammock tonight."

Charles climbed quickly down the slippery tooth, grabbed the saw and forced his way over towards his target. Jimmy hesitated a moment and then, muttering about amateurs slicing their feet off, he followed him to help. Alone on the fang I grabbed the machete and attacked the tropical gorse bush with venom. I must have cut all of an eighth of an inch into it when I heard the saw shrieking against wood. I glanced over and was surprised to see the clouds of blue exhaust gas rising up. I hadn't noticed this at all when using the saw although we had choked and spluttered on it the whole of the morning. But the smoke and the noise were the only signs that homunculi were around.

Charles and Jimmy were completely hidden in the undergrowth at the foot of the tree. Then the familiar explosive crack rang out and a Big One heeled slowly over, gathered speed and ripped its way through its smaller brethren to hit the rocks with a thud. Not bad, I thought, returning to my steel bush. The second tree was down in record time too, and I began to envision passing soup to the pasha in his hammock that night. But minutes ticked by and there was no shriek of saw biting into bark, no clouds of smoke. Charles called out that the bet was off. He'd spotted a carnivorous plant, a bright red horror with the remains of its prey sticking to its hairy lips and fronds. Science comes first he shouted, and this beauty's going into the collection box, even if it takes an hour to reach it.

It took me about an hour to get through the steel tree and at last I kicked it triumphantly over the edge into the crevasse. I climbed down off the tooth and cut my way across to Jimmy who was still wrestling with the saw among the Big Ones. I took it from him and handed him the machete. The demon team was in business again. And we stayed in business another two hours. By six o'clock we were stumbling about like the undead, mute and unseeing. We had to force our swollen fingers off the saw or the handle of the machete when it was time to change over, and then force them to grip again with the heel of the other hand . . . Just another Big One, we would tell each other. And then . . . just those two over there. Keep going or you'll sleep on your feet. Keep moving or you'll drop. In the gathering dusk we climbed wearily up the rock for a last inspection. We looked around carefully, calculating the distance and height of the nearest Big Ones. We reckoned our clearing was at least 40 feet across, with the fang jutting up 20 feet in the middle of it. We gave each other a clap on the shoulder.

Heliport Number Two was cleared.

We started back along the trail for the bivouac leaving the gas tank and the oil at the foot of the rock. Jimmy and I humped the saw along in turns of ten minutes each. Charles followed with a sack of specimens and some long, slender plants carefully carried in his hand. It was a painful business forcing our way through the ruin around the first heliport, and across Prickly Bridge, the Leap in the Dark, and all the other now familiar obstacles along our way. At the winch loading point we tied on the sack of plants, the walkie-talkie and, finally, the saw. Goodbye, you bastard. On up the gully in the near-dark, along the level bit and then we stumbled at last into the bivouac and collapsed.

After a few minutes Charles bestirred himself and twirled the handle of the telephone. "Hello, Lionel? Take the winch up, will you. The saw is on the end and a sack of plants. When you've looked at the walkie-talkie will you ring back?"

We needed that little toy. Tomorrow we aimed to cut right across the bowl to the foot of the north wall and that would be a long way from the telephone. We didn't, of course, need it to ring the corner store for a six-pack. We needed it because Charles was the leader of the expedition and by far the greater part of it was up on the surface with a busy life of its own. One major question was priorities for the helicopter, with half a dozen scientists wanting it at once, and the crew with an eye on the weather and the fuel, and it had to be settled by him alone and settled on the spot.

We got the lamps lit and the stove going and soon the remains of last night's splodge of lentils was warming in the pot with a couple of packs of powdered soup to give a bit of zip. We were a pretty contented trio. We'd done a power of work on those two heliports. Charles had made a full day's collection. Tomorrow we would get over to the other side of The Pit and see if we could find a tunnel running out to the northern escarpment of the plateau. Above all, once the work was done, our way out was secured. We'd have a breezy climb up the rope ladder and would be in on a history-making, daredevil bit of flying.

The phone tinkled and Charles picked up the handset. There were a few preliminaries and then he turned to us with a pleased grin on his face.

"They are putting me through to the pilots in Camp One."

Now we'd get a time and a date for our rescue. Charles was talking again.

"*Pablo, cómo estás, hombre* . . . how's it going?"

He listened a few moments then sat down suddenly on the ground. The moments ran on into minutes, which was odd for Charles. Then he hung up and turned to us.

"They've been working it out, the pilot says. There's nothing we can do down here to make the pickup less than suicidal.

"They can't get us out."

6

Up on the surface of the plateau, dramas of another sort where being played out: science stalking new discoveries. As it happened, Slave Labor Day in The Pit had been Botany Day on top. Julian Steyermark left his hammock early and began to force his way south from Camp One, carrying his field press; a simple folder of hickory laths and leather straps made thirty years ago in Chicago and used ever since. All it carried for the moment was sheets of newspaper, and assistant Victor Carreño cutting the trail ahead, had a sackful more. Newspapers mean a lot to botanists, but not for reading. A normal newspaper page, folded once, happens to be the perfect size to hold cut specimens and to lug about in the field. You'll be delighted to know, next time you visit a botany museum, that the stiff white sheets of paper on which the exhibits are so beautifully mounted, were not trimmed to their more or less 12 by 15 inches in size for aesthetic reasons, but because that size corresponds roughly to a folded page of the *Wampum Bugle*, or whatever journal the collector takes with him into the wilderness.

In the field he cuts his sprig of leaves and flowers and puts it between the folds of his newspaper. He must see to it that examples of both the upper and the lower side of the leaves are visible and must make the neatest arrangement possible because the field press will make that arrangement permanent. These are simple tools for a team whose object was no less than a complete inventory of everything growing on the summit of Sarisariñama, known or unknown to science.

As they moved through the forest, Julian pointed to a tree and Victor, an expert in his own right, shinned up, snipped off a branch from the top, brought it down and handed it over. Later, Julian recorded it in his notebook, with a beautiful paragraph:

> 108860 Buchenavia
>
> Tree 10 meters; flowers slightly fragrant; calyx tube creamy buff within, flesh tinged on outside; filaments whiteish; anthers pale yellow; ovary gray-sericeous; leaves oriaceous, deep green above, pale green below.

Only a painter would see so much in a tiny blossom. But not only is the passage beautiful, it describes a new species.

That number at the beginning is worth repeating. It means that Julian had collected 108,860 separately numbered plants. By the end of the expedition this was to grow to 109,880. But, as he sometimes collects several— maybe as many as five or six examples of a single plant—the number of sprigs and sprays and flowers and whatnot he has put between folds of newspaper and strapped into his press during his life is more than half a million. No one in the world can match this record. His specimen No. 1 was collected in his native Missouri in the twenties.

But there was no time for reminiscing and neither was this the scenario for it, for all around them were thousands of strange plants, up to three feet high, of the *papagayo* and *caucho* family (Euphorbiaceae) known in only one other spot in the world—the distant plateau of Yutaje in Venezuela's Amazonas State. Is this the same or a new species, thought Julian. If it is the same, how do we explain how they came to exist on two different plateaus with such an immense distance between them? No bird or wind could ever carry the seed from one to the other.

They collected half a dozen samples and the piles of folded pages grew apace. When they were eight inches or so thick they tied them in bundles and dropped them on the trail to be picked up on the way back. At night in camp the bundles would be dipped in a three-foot plastic bathtub holding a mix of one-third formaldehyde and two-thirds water. It's poisonous, it stinks, and it can burn your hands, but you have to have it. Once soaked the bundles are held up until the liquid drips out of them, then are wrapped in plastic sheets. They can stay in this package for months if necessary until they reach the laboratory where they are redried carefully in warm air chambers. Well treated, they will last for centuries, like the collections of the eighteenth-century botanist Linnaeus, still to be seen in the British Museum and the Swedish Museum of Natural History in Stockholm.

The layers of each bundle were in meticulous order and each was noted in the log with its number and description. It's the sort of punctilious docketing job for quiet, comfortable white-smocked men in laboratories, not for muddy blokes in the wet, windy jungle of Sarisariñama. But Julian was in his natural habitat. He had explored more plateaus than any of us except Billy Phelps.

While working in the Chicago Field Museum of Natural History, he was sent here in 1943 to search for sources of quinine. He had long dreamed of the Lost World region of Venezuela and his first contact with it was his ascent of the famed Roraima, the central feature of the remote land immortalized by W. H. Hudson in *Green Mansions*, and Duida, where he was the second man to reach the summit. Moreover, having got to the top of a plateau, it wasn't his custom to about-turn and hightail it back to the comfort of the city and the plaudits of the learned societies. He would stay put and work. In 1953 and 1955 he spent four months and five months on top of Chimantá, southeast of the Auyán-Tepuí, the plateau down whose vertical walls tumbles the 3,285 foot Angel Falls, the highest in the world. Finally, in 1959, he came to work permanently with the National Herbarium of Venezuela to study the flora of the country and to produce the same sort of comprehensive volumes he co-authored with Paul Standley on the flora of Guatemala. The publishing of this work began in 1945 and its compilation is still being completed by the American botanist Louis Williams. This type of study takes years. Julian says, "The work I'm doing now will still be being published after I'm dead."

But right now our star scientist was feeling very much alive as he brushed aside a shrub with his stick and peered at the treasure beneath it. It was a *navia*, a plant of the pineapple family, but with no resemblance what-

ever to the untrained eye. The navia is a star of wavy
leaves six or seven inches long, white and green, with a
yellow flower at its center. Since exploration of the
plateaus in this region began, 65 species of this plant
have been found. Was this a new one? And here's a
red-flowered shrub. Clearly it's *Raveniopsis*, but there's
something odd about it. Maybe it's new too.

The day was running on excitingly enough. But the
big find was yet to come.

If there is one thing a botanist dreams of it is finding a
new genus and family of plant. Julian was about to do so.
Ahead of him he recognized a tree similar to some he had
discovered on the plateaus of Yapacana and Auyán-
Tepuí ten years before. He had never been able to figure
out its family and neither could the specialists in this field
in Kew and the Smithsonian to whom he sent specimens.
During that decade the question became so insistent in
his mind that he went back to the Yapacana in an
arduous expedition through jungle waist-deep in water,
to get examples of the tree's fruit. Now he got Victor to
climb up and cut branches with their yellow flowers. This
would be the final clue once it was back in the world's
leading laboratories. Julian studied it. Was this the same
as the trees on the Auyán-Tepuí and Yapacana? Does
it exist anywhere else? He had that marvelous feeling.
Here was a plant in his hands and it didn't even have a
name.

The *Guinness Book of World Records* says that the smallest
orchid plant in the world is called *Notylia norae*. What it
does not say is that it was named for Nora, wife of Stalky
Dunsterville, and was found by them in 1957 in the
remote Icabaru country in the southern jungles near the

Brazilian frontier. Such exoticism is a way of life for this couple who have been probing the hills and forests here for twenty-five years pursuing their goal of illustrating and describing all the species of orchids natural to Venezuela. Stalky is now preparing his sixth volume. When it is published the series will cover about 1,000 species, and he estimates there are at least 200 more to go. The Dunstervilles, on their own or in joint expeditions with Julian, have discovered 40 new species and one new genus, which in collecting terms looks to be a remarkable record, particularly when specializing in a single plant family. But Stalky modestly brushes it off, saying that what it really shows is how little botanical work has so far been done in the tropics.

Ever ready to do their own bit to set this to rights, the Dunstervilles would rise betimes in their spartan camp near the heliport and probe in a different direction each day, tackling the thick forest and the deep crevasses of the open rock areas on their own—no machete men or porters for this doughty pair. In the days we were down the hole they found about 50 different species. Those in flower they identified on the spot. Those without a flower were collected to be carefully tended in Caracas for as long as two or even three years before the blooms appear, for until this happens the plants cannot be surely identified. This was a good collection for a limited area at only one altitude level.

Most of the plants could be reached from the ground but when necessary, Stalky, who celebrated his sixty-ninth birthday on the plateau, would shin up a tree to get them.

"Or I'd send Nora up," he says wryly.

The specimens are carried in an ordinary net shopping bag, chosen for the task because it lets air through. In camp Nora cleans up each plant, trimming off dead roots, leaves and bulbs to prevent them from rotting before they can get them back to Caracas. She cuts the living roots to about four inches long, so they will not wither in transit. They are packed, by the way, in open plastic bags which are put inside sacks and guarded devotedly from lumbering expedition riffraff such as cooks and climbers. Once the plants are cleaned they are set up on a folding canvas stool. Stalky, with pencils and greaseproof paper, sits himself on another one and the magical process of illustration begins. Many collectors have found that a camera cannot equal meticulously wrought botanical drawings of plants and their various parts. But few can do this job themselves. Stalky's drawings are beautiful things, famous for their impeccable accuracy. These drawings, done in the rough conditions of the field, are the basis for the finished pen-and-ink illustrations in his books, and are sent for final indentification of the specimens to his co-author Dr. Leslie A. Garay, curator of the Oakes Ames Orchid Herbarium of Harvard University, of which Stalky is a research associate.

It was one such drawing, perhaps the third he had ever done, that finally set him on his second career, and one so vastly different from his successful life in the wheeler-dealer international oil business. In 1953, finding he couldn't get an orchid identified in Caracas, he sent a drawing of it to Kew. The experts replied, saying how very interested they were to see a drawing of a very little known orchid that had been described by Bouché in 1848, and had never been seen since. The Dunstervilles had become interested three years earlier in the orchids many Venezuelan enthusiasts grow in their homes. They began to scout around in the forest whenever they could and soon found they were collecting specimens nobody recognized. The letter from Kew was the final proof that not only had they found an exotic field of activity but also one where work of real scientific value could be done. They were hooked. When Stalky retired from the presidency of Shell de Venezuela in 1957 he and Nora became full-time orchid hunters.

By the light of a single candle in a folding lantern slung under their tarpaulin—no ninny tents for these two veterans—Stalky writes up his notes, and rare, arcane work it is:

No. 1316   EPIDENDRUM FRAGRANS SSP. FRAGRANS

Plant.   Epiphytic, a repent or scandent, occasionally branching, subterete rhizome, clothed in generally persistent, rather formless, grey sheathing and bearing approximate or openly spaced pseudobulbs.

Pseudobulbs.   Unifoliate. Very variable, from small and subglobose to elongate, moderately compressed and stipitate forms, to 7×1.5 cm. The basal sheathing small and fugacious.

Leaves.   To 9×2.5 cm., thinly coriaceous but fairly rigid, the margins lightly revolute, the mid-nerve sulcate, dorsally indicated only by a narrow dark green line.

Inflorescence.   Terminal, from a relatively long, loose, grey sheath at the apex of the fully mature pseudobulb, the peduncle erect, short, stout, and bearing a few-flowered, very short raceme, the whole to 4 cm. Rachis lightly compressed, floral bracts insignificant; pedicellate ovaries to 2cm.

Flowers.   Sepals and petals creamy white, sepals to 20×5 mm., petals to 18×5 mm. Lip adnate to base of column, the free portion about 15×14 mm., concave, basally truncate, apically short-acuminate, white with dark purple nerves that appear dorsally in paler form. The callus is subdued, glabrous, white. Column white, stout, with a very prominent, attenuate, terminal ligule. Anther creamy yellow. Bollinia two pairs, strongly compressed, on narrow white stipes.

Habitat.   Fairly common, mainly on gnarled Bonnettia trees, in the more open parts around Brewer Camp No 1 on Sarisariñama (alt. 1,400 m.) less frequent in the more shaded or dense parts of the local dwarf forest, not present in taller forest. In fully exposed places most plants were dwarfed and depauperate but otherwise healthy. Specimens were also sent up from the bottom of the main 'hole,' alt. 1,100 m.

This is grand stuff, as resonant as a pirate's cuss. "Well, scandent me rhizome!"

And, of course, it is hardly surprising that so colorful a language had to be invented to describe what are among nature's most colorful creations. Pale jade, flamant yellow, emperor's purple, there are orchids of every hue, and they grow on plants from a quarter-inch to twenty feet tall. The flowers may be nine inches across of the sort matrons will make into exaggerated corsages. Or they may, like the Venezuelan *Phragmipidium caudatum,* have threadlike petals two feet long. Some are smaller than the head of a pin, a mere fleck of color. But under a microscope they explode into dramatically lovely flowers with three sepals, two petals, and the lip—the dazzling center piece vividly colored to attract the insects needed for pollination.

Some of us rarely see an orchid except in a florist's window, but the Dunstervilles have more than 500 growing in their house in the hills south of Caracas. They have collected them in more than 15 major expeditions including those to the Devil Mountain, or Auyán-Tepuí in the language of the Indians, where they were the first collecting party to penetrate the northern half of this 225 square-mile, 8,500-foot-high monster of a plateau; from the Sierra Avispa in the far south territory of the Amazonas near the Brazilian border; or in the Perijá region on the edge of the fierce Motilón Indian country close to the frontier with Colombia. On their five-day march into this region they came upon a human jawbone sitting on a rock near to a secret trail used by illegal Colombian immigrants on their way to seek work in Venezuela. Near this macabre find was a pile of poor, peasant women's clothing. Victims of a jaguar? Man-eating ants? The Dunstervilles learned nothing until a month later in Caracas when they read newspaper accounts of the capture of a notorious bandit responsible for 17 murders along that trail. He had recently raped and killed two women, and then, feeling peckish after these exertions, had chopped off their heads to make a stew. Picking one jawbone clean, he had laid it reverently on the boulder as a monument to a splendid lunch.

It was after their own lunch of mixed nuts and water on the fifth day that the Dunstervilles made their first major find. A tiny orchid plant with lentillike leaves clinging to a sapling. It was a new species. More interesting still was one Charles had found in the hole and sent up by winch. Its seed capsule was growing close to the roots instead of at the base of the leaf. Around Spring 1975 it may blossom and prove to be another new species. They also found in profusion an epiphytic (tree-grower) orchid which may be a subspecies new to this continent or a species new to science. But of women's bones they found none this time. Not even a tooth.

In the realm of the exotic little can vie with tropical orchids. Except tropical birds. Vivid, brilliant, varied, they had been drawing Kathy and Billy Phelps into the jungles for forty years in a career of expeditions unmatched in the story of Venezuelan exploration. In almost all these forays they penetrated regions that were unmapped and unknown. They pioneered the approach routes, befriended the Indians and opened the trails that were later followed by other scientists and naturalists. Pursuing their lifelong objective of finding out what birds exist in Venezuela and their habitats, they have quartered this beautiful land from the arid, crystal-sharp coral islands off the north coast to the glaciers of the Andes, and south through the forests to the Brazilian border.

The Phelps team has reached the summits of more plateaus than anyone else, has spent more time on the trail, and is responsible for the collection of about 1,300 of the approximately 8,500 species of birds that exist in the entire world. They have thus collected more species than are to be found in the whole of the North American continent, including the States, Canada, Alaska and Greenland. South America is, of course, as Billy disarmingly points out, the Bird Continent, and is richer in species than any other region of the earth. But this is the work of just a single man-and-wife team and is a performance recognized by their peers. In a joint Phelps Collection-New York Botanical Garden expedition in 1957 to the Cerro la Neblina, the Mountain of Mists, at Venezuela's southernmost tip, the American botanists Basset Maguire and John J. Wurdack measured the massif's highest point. They named it Pico Phelps. And that's how it appears in world atlases—Phelps' Peak. At 9,900 feet it is the highest mountain in South America outside the Andes and the Santa Marta range in Colombia. The moral is: If you work hard and successfully for forty relentless years you've still got to be very special to get a mountain named after you.

Billy, with proper scientific perspective, is more con-

cerned with other tokens of his efforts, namely the heavy
volumes containing his ornithological studies. Just one of
them with its maps, its photos, its tables, its charts, its
meticulous texts would be a proud monument for us
lesser mortals. But he has written *eight* of them, and is still
writing. Not to be outdone, Kathy produced her own
work on the birds of Venezuela, with 100 color plates
printed from her own paintings. It has run into five
editions, so far.

To catch their birds, collectors like our team use nets
and shotguns. The fine nets are strung between trees.
Light No. 12 shot is used in small-bore shotguns so that
as little damage as possible is done to the specimens. It is
difficult enough to get close to a bird to take a shot at it.
It is even more difficult to find it in the thick undergrowth afterward. For a start, they never fall in a straight
line and many of them are colored to blend with the
vegetation they fall into. When they are located, the
collector stuffs their throats with cotton or wool, to
prevent insects from crawling in, wraps them carefully in
newspaper folded in the traditional shape of a cone with
a turn down flap and puts them in his bag.

Back in camp, the skin is removed complete with
feathers, leaving the skull and the bones of the tail, feet
and wings intact. The inside of the skin is treated by the
normal preservative methods of the taxidermist and it is
then stuffed with cotton or tow. This produces a scientific
specimen ready for the museum. In the jargon of the
trade the operation is called "making skins." Still out in
the field, a label is tied to each specimen recording
essential details.

An example:

No. 72394   Sarisariñama, Feb 15, 1974
                              Altitude 4,550 feet

AMAZILIA LACTEA ZIMMERI

Total length 105mm. Iris, dark. Upper mandible,
black. Lower mandible, brown. Legs brown.
Habitat, low jungle. Collector Kathy Phelps.

The label also carries an outline sketch of the sex
organs. From the size of the gonads, the expert can
determine whether the bird was taken close to, or during
its breeding season. The weight of the birds is taken on
special sets of spring balances. They may weigh from
more than two pounds down to about one gram, or .32 of
an ounce. Birds of this size or less are hardly worth
weighing at all—a recently swallowed insect could
double the tiny creature's weight and throw the record
off by 100 percent.

The specimens are recorded in swift, unerring water-
color field sketches by Kathy before they are packed
away. They are also relieved of all parasites, some of
which live only on birds and even only on some species of
birds. They are all part of the studies, even if, as Kathy
discovered on Sarisariñama, they are tiny enough to live
in the film covering a bird's minute eye.

No one is more opposed to killing birds than the
ornithologist, but no one knows better than he that there
is no other alternative if knowledge is to advance. This is
especially true in remote, unknown areas. Specimens
must be taken even of common, easily recognized species,
otherwise there is no proof that they were in fact to be
found there. Cameras are no substitute. It is difficult
enough to photograph a bird and even pictures by
experts would probably only suffice for identification at
the species level, and not for the many different subspecies which are only identified with confidence after
scrupulous and minute examination in a museum or

laboratory. In any case, birds are never shot, inspected and thrown away. On the contrary, the carefully preserved specimens should last for centuries and still remain useful for scientific studies.

The Phelps Collection was started by Billy's father and is housed in a rambling building on the grounds of his old home in the El Paraiso district of Caracas. It is a cool colonial building with what is probably the finest ornithological library in Latin America, a card index system, and 40 tomes of meticulously collected references and notes on the specimens. The birds themselves are stored in trays inside large metal cupboards modeled on those in the Smithsonian.

But Camp One on Sarisariñama was a far cry from the order and quiet of the museum. It was a mixture of mud and mayhem. On the first day up there we had dug holes in the sandy earth and collected the evil-smelling water that seeped into them. As these sources petered out the expedition was reduced to cutting down a type of bromeliad plant, *Brocchinia acuminata,* which holds rainwater in quart-sized chambers walled with tight-fitting leaves. These plants may grow up to ten feet long and lie on the ground like some great crude earthworm, rising about four vertical feet at their leafy end. They grew in profusion around the camp site, but each day they were slaughtered by the dozen, and their heavy, ugly, limblike trunks scattered in a widening circle of desolation. Through this sloppy ruin Kathy would pick her way daintily in her high jungle boots to help out in the central slum of tarpaulins slung over pruned trees where the riffraff lived and the cooking was done, or to vanish down the trail in the bush cut by taxidermist Gilberto Pérez.

As their collection grew Billy checked them against the list of 98 specimens they had garnered on the southwest tip of the plateau seven years before and the 30-odd taken with veteran explorer Felix Cardona on the talus slopes in 1943—thirty-one years earlier. Two years before Jimmy was born. Five years before that, Billy had been on Roraima. In 1937 he spent several months exploring the immense surface of the Auyán-Tepuí, Devil Mountain. The expedition reached the 8,125-foot summit and collected 1,217 specimens from the 3,600-foot level and above. Of these, 448 specimens represented 59 species of birds of the plateau regions. His laconic report, published in the *Bulletin of the American Museum of Natural History*, says: "This sample can be considered adequate."

But it was not these branches of science that held sway at night around the cooking pots. The big question then was how the holes were formed. One night Archie, stirring a great bowl of goo, announced with authority that The Pit was a silo. That 6,000 years ago it was dug out by visitors from space. "First, they landed their ship just south of here. The fissures in the rock there were caused by the heat of retro-rockets. They built the hole as a sort of dry dock where they could make repairs to the ship's sides and then blast off to go back home. I think one of them missed the flight because he's standing behind Andrés with a . . ."

Despite himself the pilot glanced quickly over his shoulder at the dark forest and then recovered himself.

"I think we'd do better to ask someone who knows. Like a geologist," he said haughtily.

Antonio and Pablo took the cue. They said that any explanation of the presence of the holes would now have to be related to the new theory that they and other scientists were evolving on the geological history of the whole Guiana Shield. In simplest terms this theory was that the Shield had not been as stable in the past as it is in present times. Its earlier life, over a period of about 3,800,000,000 years, had been tremendously complicated. There is evidence that from 1,900,000,000 years to 1,700,000,000 years ago shallow continental seas submerged the whole area and the Shield had been in large part covered by a sequence of sediments, which eventually hardened into rock. These strata were thousands of feet thick.

When uplift pressure from below in the earth's crust began to force the Shield above the seas, major faults and fractures provoked the first stage in the formation of the isolated plateaus which were later sculptured into their present shape by erosion. Although they are about 1,700,000,000 years old, they are the *youngest* major geological event in the life of the Shield, which dates much further back into the enormity of geological time, back almost to the first cooling of earth, 4,500,000,000 years ago.

Pablo pointed upwards. "In earlier times, there were hundreds, perhaps thousands of feet of rock above the strata on which we are now sitting. This eroded over a period of millions of years until the top of the plateau came down to its present level. The uplifting movement had opened vertical fissures in the rock. Rainwater, percolating through the pores and fissures of the upper strata, found its way deeper and deeper. When it met more impermeable and unfissured strata it forced its way horizontally along the top of these strata, carving tunnels and then widening them."

"That's what formed the underground rivers?" Julian asked.

"Yes. But it has to be seen as a major process, a

network of underground rivers spreading out over one or several strata. Eventually some of the tunnels were blocked by roof-falls. At these points tremendous pressures built up and the water formed upward-spiraling whirlpools relentlessly eating away the roof until huge underground caverns were carved out. Some time after forming the caverns, the water found its way out leaving them empty. Then, possibly very recently, when erosion had brought the summit down to its present level, a diaclase system of vertical fissures developed and the roof collapsed."

"That must have been a noisy moment," said Stalky.

But there was silence in the Slum while keen brains sifted this information, seeking out its implications. Pilot Andrés was the first to deliver his findings.

"Could there be other caverns down below whose roofs have not yet collapsed? Could we be sitting over an 800-foot cave that's just waiting for its geological moment to fall in?"

There was some shifty movement. People peered in alarm at the ground and looked round for somewhere to run.

"It's possible. But the collapse of the holes of Sarisariñama must have been more or less at the same time, geologically speaking, that is," said Antonio.

Archie was going round the circle handing out hard-tack biscuits. Reaching the last man he demanded silence for an announcement.

"Well, now that you've all heard the newfangled theories of contemporary geology, I'll tell you what a venerable old greybeard in the university told me before we left. Your theory is a roof-fall; his theory is a wall-fall."

Archie explained that on Sarisariñama each hole is at

the bottom of a wide, gentle depression which concentrated the rainfall in its center where a diaclase was formed. This was a crisscross of deep vertical crevasses at right angles to each other as if a giant knife had scored several lines from north to south and several across them from east to west. If other faults or fissures should occur at yet a third angle, say from northeast to southwest across these squares, a five- or six-sided plug may be formed, roughly like a cylinder, separated from the surrounding rock. Try it with a pencil and paper, Archie said, and you'll see how. The action of the acid waters peculiar to these sandstone plateaus could, over millions of years, break down this plug by washing away the material holding the grains of rock together. This would leave a roughly circular hole with vertical walls. According to this theory, the holes should get wider as giant, vertical segments of rock round its sides are cut off by fissures and collapse into it.

Pablo was shaking his head. "I'm afraid your grey-beard is definitely off net. His idea is totally out of place in this type of formation."

But for us three down in the bottom of the hole it was very much *in* place. For one thing, there were great segments of rock round the rim, separated from the main surface of the plateau by fissures. The whole of the northwest arc was like this. We had heard the theory and had accepted its rejection by the geologists—until we got down there. Peering up at those huge looming walls above us, it was difficult to be quite so confident. The idea was, of course, the major factor in Jimmy's phobia about rock-falls, and once we heard the crash and rumble of the first one, we inevitably felt less certain. As chance would have it, Old Greybeard's theory was to be thrust into our attention again at the worst possible moment in our six days in The Pit.

7

FROM THE MOMENT WE WOKE, PEAKY AND WAN, ON DAY four in The Pit, it was clear that Charles was hatching Escape Plot No. 2. He climbed out of his hammock, wandered off to take a leak, wandered back, accepted a cookie and a dried prune for breakfast without a word or a change of expression, and without shifting his glassy-eyed gaze from some point straight ahead of him.

"He's thinking of buying an elevator," I said.

Jimmy, my partner in the Sarisariñama Heliport Construction Corp., was examining his blistered hands. "Well I'm not going to do any work on it until the maker says it will go," he said.

I stretched my aching muscles, leaned over and prodded him with a finger. He recoiled with a yelp of pain.

"I'm glad you are as sore as I am this morning, and it's not because I'm getting old," I told him.

"I don't see there's anything at all to be glad about right now and, by the way things are going neither you nor us are going to have the chance to get much older."

He looked up through the dim, green light at the bottom of the hole to where the rim shone golden with the rising sun. "That's where I want to be," he said. "Up there in the land of the living. Not down here in this tomb."

He fixed me with his eye and pointed an accusing finger. He was warming to his morning theme. "Right now in my new apartment which we have just finished making into a warm welcoming home, my wife is waking up to see to my one-month-old baby in his cheerful nursery. There's a smell of coffee coming from the kitchen and there's a fresh melon in the icebox. And I left all that to come on my first expedition with two maniacs and you get me trapped in this hole. Do you know what the date is? It's one thousand million years before the cavemen. We don't belong down here. At least *I* don't. Maybe you two weirdoes . . ."

He stopped in mid-sentence, staring wild-eyed at the pot on my knees. It contained the remains of last night's supper. Cold spaghetti, soggy with cold chicken soup.

"What are you going to do with that?" he asked, pugnaciously.

"Eat it."

"You're going to eat *that?*"

"Yes," I said, puzzled, reaching for the sugar. "It's very good if you sweeten it enough."

"Sweeten it?" he said faintly.

I handed it to him. "Here. Have half of it. Put some jam in, if you want."

Jimmy pressed himself back against the rock and raised his head.

"Jesus get me out of here," he shouted.

The echoes of his prayer died away in the jungle and the heavy, eternal silence blanketed us again. I sat thinking of the northeast arc of the wall, or what I had been able to glimpse of it from the heliports the day before. Maybe there was a route up it somewhere, but the bottom hundred feet or so seemed blank and impossible. Jimmy was reading my thoughts.

"If you are working out a route up that part of the wall you were looking at yesterday, you can count me out. I'd rather die in one piece than. . . ."

A slight noise made us glance over at Charles. The oracle was stirring. His eyes came slowly into focus as he came out of his trance, and shone upon us with a gleam of triumph. "We are going to get out of here," he intoned with a seraphic smile.

Jimmy and I examined him in silence. We'd heard this tale before, and we had paid for it dearly just yesterday. But nothing gives you an open mind like being trapped down a prehistoric pit.

"Okay Charles. Just tell us how?"

"By winch," he said.

Now there are more crisis factors in taking a ride by winch than you'd believe possible. Instinctively, we looked over to the 750-foot overhanging wall down which the winch rope hung freely and giddily to the rock platform on the descent route to the cave. It was intimidating just to look at it, and as for riding up it tied under that conical cage. . . .

We turned to interrogate our resident wiseguy. First, there was the question of the rope. The winch rope was in fact two ropes joined together. The bottom one—the green one—was only a quarter-inch thick and had a stiff center core. God knows what it was originally made for. The top one was a white, silky-soft affair about one-half-inch thick. What knot would you use to join two ropes where not only is one thin and the other thick, but one is stiff and the other pliant? I reckoned I might settle for a double fisherman's bend and then a tarbuck tied with the free end of the soft rope round the hard one. But the winch crew were not climbers. They weren't even sailors. What horrible, primitive, deadly, unsafe ribbon-bow had they used? And how many more loads would it take before it popped undone? I explained this in rather more lurid terms, and backed by Jimmy, indicated we'd rather embrace Cuyakiare's grandmother than ride up three feet on that rope.

"Aha!" said Our Leader. "But I'm going to telephone the surface to ask Luís Armando to fly to Caracas to get a new rope."

Talk about thinking big. Luís Armando's plane was at Sta. María. The whole helicopter schedule would have to be thrown out of kilter to get him there. His takeoff and landing problems were hair-raising, and getting to Caracas and back was a 1,000 mile-flight. Jimmy and I were properly impressed, but not speechless.

"Hold it!" I said as Charles reached for the telephone. "Where are we going to get a single rope 800-feet long? And I mean a rope, not some manky ball of string?"

"From the factory. The owner is a patient of mine." He reached for the telephone again.

"Hold it! He doesn't make kernmantel climbing rope. He'll have to give us that three-ply 11 mm., medium tensile nylon stuff instead."

Charles made another attempt to get at the phone.

"Wait! The rope must be . . ."

He held up his hand for peace and said we'd better get the airlift moving first and talk about the rope afterwards. He would call the factory when Luís Armando was on his way. We couldn't risk the weather closing in and stopping him from leaving. This made good sense and I shut up as he twirled the handle on the telephone and asked the surface team to put him through to Luís Armando in Camp One. Charles explained the mission and Luís agreed at once to make the flight. There was only one problem. The helicopter had already gone to Sta. María, for fuel, and the clouds were closing in.

Charles replaced the receiver in its box, and Jimmy spoke up. "Are you two really serious about going up that wall on the winch?"

Yes, we said. Why not?

"Suppose there's a fault in the new rope. If it breaks when you're only 50 feet up you're dead, and the wall is more than fifteen times higher than that. And what if the winch motor fails, or we get stuck on the final overhang?"

In truth, we were not happy about these questions ourselves. You don't fool around with a 750-foot drop. As we constantly reminded each other, death is irrevocable, and should not be risked more than twice before breakfast. We decided to ask the surface team what they thought and Jimmy picked up the telephone.

"Hello, Lionel. You heard the plan when Charles talked to Luís Armando? What do you think?"

He listened in silence then hung up.

"They don't like it. They have set everything up as best they know how, but say it's one thing to haul up specimens and another to have a live human on the end of the rope who will be scraping up the rock on the final overhangs in a position where they cannot see him. It's a responsibility they were not prepared for."

Things were getting rosier by the minute. We relapsed again into hard thinking. There had to be an answer. Suddenly I perked up.

"What about a safety rope? We have two 200-footers down here. I'll go up the route as far as the next to last ledge and get the third and most of the fourth down off the wall. Knotted together they would just about be long enough. Federico could look after the winch as we are pulled up on the new rope, and Lionel, who's a big chap, could take in the red rope which we'd have tied to our harness. If the winch or the winch rope breaks he and one of the Indians can hold us and, at a pinch, even lower us down again."

Charles jumped up excitedly. "That's it! That's the answer! With a safety rope it would be a terrific ride. I'll film every inch of it. Can you imagine panning right round the walls? Can you imagine shots taken looking straight down between your feet? Wow!"

I was delighted too. It would be a sensational trip. Absolutely terrifying. Astronauts get winched up out of the ocean into a helicopter, but only 30 feet or so, not 800 feet. Construction workers and quarrymen might ride up a certain distance in a skip or steel bucket. But we'd be tied under the iron cone, hanging free. Wow again.

Schmucks! We ought to have known that nothing is so simple in The Pit.

In the early afternoon we got the encouraging news that the weather had finally allowed the helicopter to get back to Camp One and to take off again with Luís Armando. The improved conditions also permitted

Lionel to get through to Caracas on his HF radio transmitter. And that's when our troubles began.

It took a long time to track down the owner of the rope factory and we were getting more despondent by the minute. When at last he came on the line we gathered eagerly round the telephone. From this call would come our lifeline. In a low, intense voice Charles explained our predicament to an astounded and clearly bewildered listener. Trapped in a pit? Five hundred miles away? You want to be hauled out on an 800-foot rope?

"Yes."

"I'm afraid we don't have a rope that long in stock."

There was a pause. Then he launched into a business report. Due to the Arab oil embargo there was a shortage of inputs to the petrochemicals industry and thus a scarcity of products. Particularly hard hit was the textile industry and others requiring artificial fibers, such as rope factories. I smirked around The Pit. Get a load of that, Cuyakiare.

"Well?" said Charles.

"Well, I'm sorry. I can't make one."

Our spirits slumped then leaped as his voice came through again.

"If it's an emergency, I could get such a rope rushed over here from England."

"That's great. How long would it take?"

"Oh, I'd say about two weeks."

It took us a few minutes to get over that one. And then Charles was twirling the handle again. This time he got through to an old friend of ours in the Centro, Julio Lescarboura, who immediately jumped into his car to comb Caracas for a rope. We were left with a variety of emotions, dejection because the plan seemed to be falling through, embarrassment that we had sent Luís Armando off on a hazardous trip for a rope that didn't exist, faint but irrepressible hope that Julio would dig something out from a sports shop or ship suppliers or whatever. We fiddled fretfully around in the bivouac unable to stray far in case a common decision was needed. Then the phone rang after about only thirty minutes and Julio reported he'd found a rope of the same make as our trusted red climbing ropes.

"How long is it?"

"Nine hundred sixty feet. It's in the warehouse of a sports shop and they can dig it out in ten minutes or so."

We whooped with delight.

"Terrific. How much is it, by the way?"

"Five thousand bolivars."

We looked at each other in stricken silence. That was $1,160. Charles told Julio to call back in ten minutes and we launched into a raging debate. It really is remarkable what a scar money has made on the minds of all of us. Our first reaction was to recoil from that quantity of what were no more than printed pieces of paper even though the lives of three men and the future of three families were balanced against it. Of course we finally had to accept that cost, but even so we had to justify it to ourselves by saying it would provide us and the Natural Sciences Society with enough rope for several more expeditions.

The field telephone tinkled and Charles told Julio to go ahead and buy the rope and get it to the airport to meet Luís Armando. We leaned back against the rock aghast at our financial temerity. But in a few moments we had recovered and decided that as we were proving to be such expensive expeditionaries we'd better get cracking. Our escape was settled, so let's get some work done on the trail towards the north wall before dark. We jumped up in high fettle, grabbing the machete, the rope and the caving gear. Jimmy and I were already moving along the ledge when we heard the telephone tinkle and Charles talking. We stopped and then turned back. Charles had replaced the receiver but was still sitting next to it. He looked up as we reached him.

"Julio got the rope out into the street and measured it. It's only 650 feet long."

We dropped our gear and I looked round the hole. I felt sure this superbly serene but malevolent pit was watching our antics with what had to be contempt. I was beginning to get cross with it.

"Right. We get that rope, join one red rope to it, use that combination for the winch rope and the green and white combination as a safety rope. It will be what we call psychological security," I said.

Charles understood immediately and agreed. Jimmy was not so sure. I explained that when the new rope arrived it would be lowered to us still coiled. We would examine every inch of it then join it with loving care to one of our red ropes and send it up again. The winch crew would take the green and white ropes off the winch and replace them with the new combination. The green and white ropes would be lowered separately, as the telephone cable had been lowered, and, even if dubious, would be something we could pretend was a safety rope.

Jimmy shook his head in resignation. "Well, there's one thing I'll tell you now. I'm not going on any more expeditions with you two madmen. This is the last time."

"It'll be the last time for all of us if we don't get out of here."

Charles picked up the telephone again and got through to the Centro. "Tell Julio to get that rope." He banged down the receiver with a gesture of finality.

"Well, that's that. I'm glad we've got that business settled."

That was a very careless remark to make in The Pit. As we would soon find out.

Lionel rang us to say he was lowering the recharged walkie-talkie which would be essential for the escape operation. We left the bivouac site immediately for the landing platform on the way down to the cave. At the top of the gully we looked up to see the load coming down over the overhangs. As it drew nearer I caught sight of the knot about 150 feet above it, clearly visible against a dark patch of rock. I blinked in astonishment. It was spinning like a propeller. Must have something to do with the way the three-ply rope is running over the drum on the winch, I thought. But, of course, that was not the only thing I was thinking.

"*Godalmighty!*" I shouted, to get the proper effect. "Look at that knot twirling."

They looked.

"Well?" said Charles.

"Well, for a start it means the safety rope will tangle with the winch rope . . ."

"And if we went up there spinning at that speed we would be unconscious or sick before we got to the overhangs," said Jimmy, grasping the situation.

Charles seized us both by our overalls. "You mothers. The hell with the new ropes, and safety ropes, I'm sick of it. I'm going up the winch as it is. On these green and white ropes."

He shot off down the gully as fast as he could climb and we tore after him ready to tie him up for the loony bin. By the time we got to the platform the load had arrived. Charles unhooked it and took the walkie-talkie out of the sack. Without pause he started to load it up again with rocks, muttering to himself.

"Nothing wrong with the damn ropes. I'll send one test load up then I'm going up myself. If we fill this sack it will weigh about 180 pounds. That's heavier than any of us. Safety rope? That's for ninnies. . . ."

He was working away like a robot and after a moment we joined in. The experiment intrigued us, to say the least. If Charles made good his threat and went up on that hair-raising ride, we'd be committed to do the same. From being a thrilling ascent with the comforting feel of a new rope and a safety rope as well, it had become a quite desperate prospect of ten agonizing minutes with not a damn thing you could do about it if that sudden split-second moment came when the grip of the rope around your ribs went slack and you began the long fall into space. But we went on stuffing rocks into the sack like condemned men forced to test their own scaffold, with that ever-clinging belief that surely the moment for its use would never come. Finally we hefted it solemnly one by one and pronounced it equal to the weight of a sturdy cadaver. Charles pushed the speak button on the walkie-talkie.

"We've put about 180 pounds of rocks on the rope. If it comes up okay we will follow. *Out.*"

Lionel's voice crackled back immediately.

"Out, be damned. What d'you mean, 180 pounds? What d'you think we've got up here, a crane? And do you really expect me to press the start button with a man on the end of that rotten rope?"

"C'mon, Lionel, let's just see how this load goes up. Then we'll talk about it."

"Well, okay. But get the hell away out from under it."

The familiar murmuring started up above and we could hear the change in its note as the weight of the load came on the line. Then it was rising slowly, the sack stretching and bulging. We scrambled up the gully as if Cuyakiare's old man were after us, and stood spellbound to watch the rest of the ascent. We watched it all the way up to the final overhangs and then lost it as it scraped up the final rocks. The murmur stopped and we looked at each other uncertainly. There was a shout from above. It was faint, but there was no mistaking its urgency. Charles switched the radio on and it blared at us:

". . . your lives! Run for your lives! The coupling has slipped. Federico is cutting the load free. Get the hell out. . . ."

We jerked our heads up and saw the sack just clearing the overhangs and starting on its awful fall, silent at first and then with a rushing crescendo of noise, swollen by The Pit into a fearful roar. We watched it all the way down. Then it flashed past us to hit the rock platform and explode into dust with an appalling bang.

Pow. End of Escape Plan Two.

But we weren't the only ones in trouble at that moment. Luís Armando was having an epic of his own. For a start, you would have to see the airstrip at Sta. María de Erebato to understand how many grey hairs one takeoff can cost a pilot. It is lined on each side with trees. Worse, there are trees and an abrupt line of hills at each end. The surface is deep grass. In the middle there is a bump so pronounced that when you begin the takeoff you cannot see the end of the strip. You take the bump at about 55 m.p.h. and then, so help me, you have to *turn* slightly left, and rush down the other side as fast as you can go. This descent is so steep that you cannot stop once you have started down it. Either you take off first time or you've had it. Moreover Luís Armando's plane was a two-engine Beechcraft Baron, which is not made for rough jungle airstrips. It needs a higher takeoff speed than it is possible to reach on this particular strip. It only just gets off the ground and over the trees. Any failure in one engine would be the end. Just to add to the problem the takeoff and landing can be made only in opposite directions regardless of wind, because you cannot take off uphill or land downhill on so short a strip with so tricky an approach and exit. On this particular day, Luís Armando had to take off with a tail wind. A fine start to a difficult flight.

He gunned his engines to lift up over the hills out of this death-trap and turned north. But there was no chance to relax. The time was 4:00 P.M. It was almost 500 miles to Caracas and he had to be there before dark, at the latest 6:15 P.M. in order to be allowed to land in La Carlota airfield which is in the middle of the city. With his companion, cameraman Gustavo Chami, he made some swift calculations. Normally he would fly to Ciudad Bolivar on the Orinoco River, about one hour away, refuel and then follow the air routes, served with radio direction aids, north towards the coast and then west to Caracas. Instead, because time was short, our pilot decided to fly 60 degrees northwest. This would take him in a direct line to Caracas, over virtually uninhabited jungle and plains for about 400 of the 500 miles of the route, without radio bearings until the final 30 minutes of approach.

But this was not all. Luís Armando's fuel gauge told him he had very little margin for error. The fact that he had to fly at gas-gobbling full speed to get to Caracas before dark reduced the margin still further. His calculations showed him that the crucial factor was the wind. He would have to make constant position, time, speed, and distance checks to determine its strength and direction. If he met with headwinds he would not reach Caracas. He would have to return to the normal air routes and seek out the nearest landing strip. The element of risk was clear, but Luís Armando was a meticulous aviator with long hours of flying time behind him. He knew the difference between risk and calculated risk. There was nothing in the first 150 miles of featureless jungle to pinpoint his position. Not until he reached the Orinoco after almost an hour's flying could he make his first check. He ran through the factors speed, distance, time. True airspeed was 230 m.p.h. Speed over the ground was about 260 m.p.h. He nudged Gustavo:

"We have a good following wind. If it keeps blowing like this, we'll make it."

You won't believe it, but Charles was brooding again. We were still at the top of the gully spitting out dust and shreds of sacking, and there he was staring into the middle distance, eyes glazed, pulse probably down to about 42.

"Your brother is on another trip, Jimmy," I said.

Jimmy took a deep breath and reached forward to push Charles over the edge. But I handed him the walkie-talkie and suggested he get through as fast as he could to stop Julio buying that 650 feet of expensive rope.

Jimmy pushed the button.

"Hello, hole to Camp Two . . ."

Charles was obviously deep into alpha brainwaves.

I tapped gently on his helmet with my knuckles. "Anything under here, or have you gone numb?"

"I wonder where they are," he said, to himself.

"He doesn't know where they are," I reported to Jimmy.

"Where what are?" he asked.

"The ladders," breathed Charles.

Ladders? *Ladders?* Visions of the old man climbing up to fix the screen on the bedroom window. Or some bloke in his orchard going up one with a basket to pick plums. Maybe it's snakes and ladders.

"They came over from Spain and we got them out of customs. That much I remember," Charles was muttering.

Jimmy and I stared at each other. "We'll have to lock him up," I said sadly.

But Charles suddenly jerked into life. "Quick, let's get up to the telephone. They're under the sideboard by the kitchen door." He tore off up the route and along the level ledge to the bivouac, whipped the lid off the box and twirled the handle like an eggbeater.

"Camp Two? Lionel? *Mira, hombre,* get the Centro and ask them to ring María-Mercedes at my house, will you?" He was whistling with impatience and drumming his fingers on his knee.

I looked over at the eastern arc of The Pit. The sun, long gone from our sight, was already lighting only the top of the wall. It was near to setting. Luís Armando must be almost in Caracas. I glanced at Jimmy who was on one knee in front of his brother rapidly flipping his hands over, palms uppermost, in that marvelously expressive Latin gesture of interrogation. I joined him and, seeing he had the attention of us both, Charles spoke up.

"What do you think of electron wire caving ladders to get up the final 200 feet and over the cornice?"

So this was Escape Plan Three.

We considered him with deep suspicion for a long moment but there was, in truth, no built-in trap that was immediately apparent. Electron ladders, invented by the great French speleologist Robert de Joly, are made of thin steel wires with four-inch-wide tubular aluminum rungs. They are light, but strong.

We began to work it out.

"How long are the ladders?"

"Well, I ordered two from Madrid. Ninety feet each. They arrived but I never had time to open the package. So I haven't checked the length or even whether there are in fact two of them."

"Okay. If they are both as you ordered, they will almost reach the big ledge. If they are shorter, we'll have to jumar up until we reach the bottom rung and then transfer from the rope to the ladder," I said.

"And if there's only one?" asked Jimmy.

"Then we'll have to jumar a hundred feet or so to reach it."

We thought this over a minute and then discussed the cornice. The ladders would not bury themselves in it. But when we put our weight on them, we might break off a section of the cornice and bring down a few hundred tons of rock and earth on our heads. As I would be going up first that was really my problem, and like a good mountaineer I decided to forget it until I got there.

We remembered then that we had got our sequence wrong, and that before our tussle with the climb there was the very real problem of whether the surface team could find the spot from which the ladders should be lowered. First, they would have to cut several hundred yards round the rim of The Pit and then find our white rope in the thick undergrowth and follow it to where it was knotted to the first red rope and follow that to the edge. They would have to find a solid anchor for the ladders, attach them with a secure knot and, meantime, make sure they didn't fall through the cornice while they were at it.

The more we discussed this third plan, the more problems we uncovered, and as the question marks piled up we resolved not to set too much store by it. The Pit had sore ribs laughing at us as it was.

Then Jimmy reminded us that it would have a real bellylaugh if we didn't get through to Caracas and check just what there was under that sideboard. In any case, we'd been down there four days now, and after the string of failures at getting out, we were beginning to feel the strain. We had to know.

Charles picked up the telephone and twirled the handle.

"Hole to Camp Two. Did you get through?"

"No."

"What d'you mean 'no'?"

"I mean 'no.' María wasn't at home when we tried so I didn't bother the Centro again." He paused a moment and then: "Was it anything important?"

That was the most stunning thing we'd heard in 80 hours in The Pit. Charles shook himself and spoke hoarsely into the phone.

"Camp Two? We *have* to get through *now*, or we'll all go bananas."

"Well, I'm sorry, fellows. The weather has closed in. There's no communication with Caracas."

8

When your life is at stake, you're wide open to new ideas. In the morning of the fifth day in The Pit, I became a radio buff. I wanted to get into Communications, the globe-circling link of peoples and nations. I wanted to speak into the ether and listen to far-off voices. Especially Luís Armando's, telling us just what had been found under that piece of furniture by the kitchen door.

Charles and Jimmy woke up and stared sleepily round until it clicked. Yes, we're still down here. They blinked their eyes and I could see them zeroing in on the unknown factor of the day. The ladders. Did Luís Armando get the ladders? Are *we* going to get the ladders? How long will they be? Unanswered questions like these generate enough nervous energy to run a washing machine, we found. We felt like human accumulators and if it hadn't been damp down there we'd have sparked and crackled like Frankenstein's monster plugged into the lightning rod. We had to find out or blow a fuse.

I looked up. The dawn marvel, that half-circle of gold at the rim, shone down through the translucent tourmaline air. It was too early. Suppose Luís Armando took off at seven and followed the normal air route to Ciudad Bolivar, and then south. It would be about 9:30 before he got close to Sta. María. Three and a half hours to wait for news.

But Charles was crawling towards the telephone. He knows something I don't know, I thought, and with the fervor of the new convert I sought enlightenment.

"What's HF?" I asked, stalling him.

"HF? It's high frequency."

"And VHF?"

"Very high frequency, you boob," he said, turning to the telephone.

Suppose he and Jimmy get hit on the head by a rock fall or eaten by a dinosaur under the north wall. I have to know about these radio things. Charles had his hand on the twiddler.

"Wait! Which is what? With which will Lionel talk to the airplane?"

Charles was puzzled by this persistence. But he answered patiently. He must have mistaken the neophyte's gleam in my eye for one of looming madness. A man on the brink.

"HF is for long distance. But Lionel won't use that because his frequency can't be picked up by the plane. He can use the VHF when he's closer, if there are no storms around or radio traffic which is more powerful."

"So why do you want to telephone? He's nowhere near close."

"*Carajo!* You tell him, Jimmy." He turned, lifted the

lid of the box, took out the telephone and twirled the handle.

Jimmy said Lionel would radiotelephone the Centro and give them the message for Luís Armando. The Centro would telephone the control tower at La Carlota in Caracas, or at Maturin or Ciudad Bolivar along the route the plane would follow. Whichever of these three Luís Armando was close to could pass the message to him by radio, get the answer and telephone it to the Centro who would relay it to Lionel. Simple. I would have to go to night school if I ever got out of this. And what message would be zigzagged back and forth across the country by this wondrous network? And what would all those air controllers think about it? How the hell, they will wonder, does a Beechcraft Baron carry *ladders?* Who cares what they think? *Was* it carrying ladders?

"Lionel, can you find out if he's got the ladders?" Charles' voice broke in on Jimmy's lesson. He must have been listening. We made a halfhearted breakfast, never taking our eyes off the telephone. But it didn't ring. Finally Jimmy twirled the handle and spoke to Lionel. Then he listened a moment and hung up, his dynamo humming.

"Nothing doing. He says the weather's closed in—for the day, by the look of it."

The voltage crept upwards until I sparked. I grabbed my helmet and the jumars and jumped up.

"Just in case we get those ladders and have to climb out in a hurry, I'm going up the route to straighten it out and clean it up. Don't eat all the trifle."

"Do you think you should go up there alone?" Jimmy asked.

"No, but I'll be glad to get away from you two."

"Okay, bloody *inglés.* We'll collect what specimens we

can from round the bivouac and keep near the telephone. Take care," said Charles.

I dropped down into the undergrowth and circled round the water point expecting to pick up our trail immediately. But I drew a blank. I thought that after the passage of three men dragging five heavy duffel bags the way would be wide and clear. Not on your life. In four days the jungle had swallowed it, it seemed. I pushed into the forest beyond the water point and then cut at right angles towards the foot of the wall on a course that had to take me across the line we had followed. Nothing doing. Finally I found my way back to the water point and found the other two stalking something or other among the rocks.

"I can't find the way," I complained, grumpy and cross that they'd seen me coming back.

"Can't find the trail? It's easy. Straight over there," Jimmy said, pointing confidently into the jungle. I pushed into it again and after several more false starts caught sight of the track when, by chance, I turned and looked back along it. From this direction it was easier to see because the leaves on the ground were sloping away from me with many of their undersides showing. There were, of course, no slashed stems to guide me because on the descent the machete had been in one of the packs. I went back along the trail to its start and put this to rights. Slash. Cut. Chop. There would be no mistake this time.

Without those damned heavy packs and without the pressure of having to find a route and a bivouac before dark, I pushed happily on, in and out the gullies, up the little walls and around the foot of rock buttresses, marveling at how we ever found our way through such a maze with our loads. At length I reached the first chimney, left the machete carefully on a ledge and started climbing. Not bad at all, I thought. We really had done well to come down that way at such speed. The chimney widened out after about 70 feet and I climbed the rest of the way up its right-hand wall to its top. Two very steep muddy gullies led up from here and I took the left one, climbing carefully so as not to disturb anything until I was sure it was the right way. As at the bottom, the fewer false trails I left the better. Once I saw that it was on the route I climbed back down again and then scrabbled up like a beginner, making as much mess as I could as a signpost. I soon bumped into vertical rock and cut right along a ledge. At the end of it was a mossy wall about ten feet high followed by a gully. And then, exultantly, I caught sight of the end of the muddy rope dangling down a dark chimney above me. Now for the tussle.

I clipped one jumar on the rope and went up, using it as a handhold. It was 90 feet long, half a new red rope we had cut on the way down to protect this section. It was tied at the top of the chimney to a tree. Somewhere above, the other half would be hanging down an unclimbable rock wall. All I had to do was find it.

A hundred feet up an almost perpendicular stretch of mud and loose rock, which I christened Rotten Roger, there it was. A 30-foot smooth cliff overhung by a cornice of earth and thick spear grass like a barbed wire fence crowning a brewery wall. Brewery Wall? That will be its name from now on, I decided. This time I clipped on both jumars, put my feet in the loops and started up. It was a hard struggle and there was no cold beer at the top. Just a ledge leading to the right under the two final walls. It was generous and we had already named it on the way down—Plaza Bolivar. This is the name of the principal square in every town and village in Venezuela, so called to commemorate Simon Bolivar, the Liberator. From the outside edge of the ledge I could look up to the final huge overhang, the monster that could seal our fate. It was clearly as dangerous and unstable as a cornice of rotten ice in high mountains, and it would be as merciless if it fell, hundreds of tons of heavy, hard blocks that would knock us 600 feet to the bottom and bury us there. This prospect was bad enough, but it was about to get worse. From curiosity, I went to the far end of the ledge and found I was on a corner jutting out into space. I edged onto little holds in the rock and looked round it. I wished I hadn't. I was going to be the bearer of evil tidings when I got back to the bivouac.

The news was not too good from other parts, either. About when I was leaving the Plaza to return to the bottom Luís Armando was approaching Sta. María from the north. Up in Camp One Lionel and Federico were using every bit of skill they could muster. They could hear Luís Armando's voice but could not get through to him. Their receiver crackled for the tenth time and a voice came through the static with a message like a poem.

"Sari Uno, Kilo Tango Sierra te llama. Cambio." Sari One, Kilo Tango Sierra calling you. Over."

KTS is the registered code for the plane. Sari, of course, was our own code for Sarisariñama.

Lionel put his mouth close to the mike:

"Adelante, Kilo Tango Sierra. Sari Uno te escucha. Cambio." Go ahead, Kilo Tango Sierra. Sari One is listening. Over.

But the same message came back from the plane.

Clearly they were not receiving. Lionel and Federico stumbled over the logs in the middle of the clearing which formed the heliport. The extra space did not amount to much, but it was worth a try. They tried again and again. But they never got through. Then came the voice again.

"Sari Uno, if you can hear me . . . I am about 50 miles north of Sta. María. I am descending visually but have thunder conditions ahead. I estimate I should be over Sta. María in fifteen minutes. Over."

The message was repeated over and over and in each break the surface team tried to get through to him. "Kilo Tango Sierra, have you got the ladders? Have you got . . ."

But the voice came back again, dramatic in its steadiness among the crackling static. "Sari Uno. Kilo Tango Sierra. Weather closing in completely over Sta. María. Thunder conditions down to ground level. Must return to Ciudad Bolivar. I will gas up there and try again. Over."

And Lionel, frantic, yelling back. "Kilo, have you got the ladders? Have you . . ."

"Sari Uno. Kilo Tango Sierra. Turning back now. So long . . . so long."

Back in the bivouac the air was electric. Charles and Jimmy had done some painstaking work, collecting, filming and making notes. They had been biting their nails too, and were about down to the wrists by the time I arrived.

"Any news?" I said.

"Nothing. The weather has closed in."

There was no bright circle of sky above us today. The

clouds were down to the surface of the plateau and formed a thick grey lid over the hole. In the southwest arc the mist was pouring over the edge and billowing down the wall. The flat, dead light and the feeling of being shut in did nothing for our morale and I decided it was not the moment for my morbid tidings.

"Well, you'll be glad to know the route is in fine shape. I moved the rope over on the wall where the knot is, so we can climb fifteen feet before using the jumars. And you'll be surprised at how your spirits lift as you gain height. It's probably psychological, but it feels physical. Like a shot of navy rum."

Charles jumped to his feet with a sudden decision and grabbed his pack.

"We have to cross to the north wall. Let's go now. They can tell us about the ladders when we get back tonight. We can't sit around anymore like this."

We moved off in high fettle along the ledge but as soon as we started climbing down towards the cave our verve evaporated and the familiar, numbing lethargy set in. We turned left under the winch loading point and slanted down to that horrible trail towards the heliports in the center of the bowl. Just beyond this point the jungle is at its thickest. North of the Fang we were entirely enveloped. The rim of the hole had become invisible. We were forced to cut what was virtually a tunnel through the undergrowth, and a tunnel, moreover, that wound up and down and around the huge boulders and for one stretch of 20 feet, beneath them.

At length we emerged from this terrain onto an upward slope where splendid great trees grew, over 100 feet high, soaring upwards not from strangling undergrowth but a deep vast carpet of dry leaves, brown as autumn. We found the top of the slope formed the spine of a hogback ridge and from it the ground dipped sharply down to meet the bottom of the wall in a sort of gutter along which ran a line of black, gaping holes. Above, the wall overhung us by over 30 yards filling us with an appalling sense of exposure to stone falls and even an image of imminent collapse. We tore our eyes away from it and forced ourselves down the slope even deeper under its beetling presence. This was a fearful place and for Jimmy, who was less used to awesome rock scenery, it was all but paralyzing. How fitting then, that The Pit should claim him as a victim.

As we slid down closer to the holes we began to sink deeper into the leaves. First to the waist, then to the chest. Our sense of insecurity became more acute by the minute. What the hell were we treading on? What was beneath this crackling, tinder-dry surface which had never received rain in all its history? Spluttering as the fine dust caught in my throat I glanced round at Charles and then at . . . Jimmy? Where's Jimmy?

In an instant we were plunging across the slope like surprised seals floundering frantic towards the sea. Our wild butterfly stroke brought us to the edge of a sudden crater. It was eight feet across and about six feet deep. It was shaped like a funnel whose walls were crumbling and sliding down to the hole in the center. In that hole was Jimmy's head and he was already choking and puking with the dust and leaf rubble creeping up his neck as he settled deeper into it. This shocking sight froze us a split second, enough time to work out that he'd broken through the layer of leaves and earth into a crevasse, that he had his two arms braced across it to stop himself from falling in, that he couldn't raise them to brush away the leaves that were choking him, that if we went down after him we might push him into it and would certainly bury him, that if we slid down-slope of him to break down the wall of the funnel we might be over the line of the crevasse and would fall into it ourselves . . . At that moment a little avalanche ran down the funnel and covered his head. That was enough for Charles.

"Grab my feet!" he yelled, and dived headfirst into the hole. He dug one arm into the stifling mass to seize his brother by the collar and with the other flailed away at the leaves to allow the victim to breathe. So far so good, but now what? I was lying flat behind Charles hanging like grim death onto the single foot I had managed to seize. Charles was head down in the funnel digging like a terrier, Jimmy was gasping for his life and bicycling with his legs over the empty air of the black depths beneath him. It was like one of those far-out balancing tricks in the circus, a trick we couldn't keep up for more than a few minutes. I had 90 feet of rope coiled across my chest and over one shoulder. By a series of convulsions I flipped it over my head and got it in front of my face. I had to keep taking my right hand from Charles's ankle, a half second at a time, to get one end loose, round his leg and tied one-handed with a bowline. Each time I took the hand off his leg he dropped a little down the hole and let out a roar of alarm. I was pretty alarmed myself as I slid backwards, releasing his legs altogether and taking his weight on the rope. I was getting quite good at this backwards snake crawl, a movement I had learned on the cornice five days before. Eventually I got to a tree, put two turns of rope round it and led it rapidly back to the funnel to drop it down with a loop on the end to Jimmy.

In a flash Jimmy grabbed it and heaved himself up until he could get a foothold on the edge of the crevasse. Charles, red-faced, reversed his batlike position so that he

too was straddling the crevasse. After a few moments of
hard breathing from all three of us, they pulled themselves out of the funnel. The one-handed bowline, which
saved Jimmy, by the way, had been shown to me, a
white-kneed second lieutenant, twenty-six years before by
a Sergeant-Major of the Duke of Cornwall's Light
Infantry in Mogadishu, Somaliland. He was not a
climber; he was a former shipyard worker from a North
Sea port. You couldn't get further from a hole on
Sarisariñama if you tried.

We decided not to be put off by this escapade. We had
to find that underground river and this might be our last
chance. Using the rope we slid down to the gutter where
the slope joined the foot of the wall and began testing the
depth of the dark fissures by dropping stones into them.
We tried about three, climbing carefully over the huge
blocks of stone that separated them and glancing every
other second up the slope of leaves in case it decided to
landslide on top of our heads. At last we found a real
horror, a narrow shaft that dropped with uncompromising verticality into midnight gloom. Just what we were
looking for. I nudged Jimmy. "You're the expert. Why
don't you nip down this one?"

Still pale, he swore softly under his breath, dug into the
pack on his brother's back, fished out the caver's
headlamp and held it out. "Either one of you two is
welcome to this," he said.

And in truth it didn't matter much. Charles and I took
it in turns to slide, climb, wriggle or squirm far enough
down these shafts to be able to beam our lamps down to
find they came to a dead end. Five, six, seven of them.
The day was creeping on and already the light was losing

its green tinge and turning grey. Five o'clock. Eight, nine holes. And still leading to a dead end about 40 feet down. Six o'clock. It would be lunacy to attempt that trail back to the bivouac in the dark. One more, said Charles. Let's just try one more. And we did. And then it was over. Ten holes and all of them blind. Ten struggles as exhausting for the two heaving on the rope as for the man clawing his way upwards on the end of it. In that moment we knew our dream of wading along that mysterious underground river into the depths of the plateau was over.

It was a lottery and we'd lost. The odds were thousands to one against us. We might spend another month down there probing the crevasses until we hit the lucky one that would take us down and down into the heart of the mountain and its secret galleries and tunnels. On the other hand, we might find it on our very next try. Very well, we're coming back one day with fewer tasks and more time. But right now there was not a moment to lose. We scrambled back up the slope in the twilight and with Charles in the lead, fled along the trail at maniac speed. We crossed The Pit in 30 minutes flat and reached the bivouac in the dark. Charles went straight for the telephone and this time there were no preliminaries.

"The ladders?" he snapped. He listened for a few moments, said he'd call back later, and hung up.

"They're here. Up at the winch site. And there's two of them."

We said nothing for a while. We were all thinking the same thing—that in a way the ladders brought as many problems as they were supposed to solve. Their arrival was only the end of a phase, not the end of the story. In that sense it was no more than the preliminary to the final chapter the end of which would either be escape or a long fall in a cloud of rocks and rubble. The thought was disturbing, but we followed it up.

"So what's the next stage?" said Jimmy.

"Well, Pablo Colvee and the two Makiritare cut a trail almost as far as the point where we dropped out the helicopter," said Charles.

"So all they have to do is follow and then find our first rope?"

"That's not going to be easy. In fact, it's a shot in the dark."

I joined in here. Doleful Joe, bearer of bright tidings. The Christian thing would have been to keep the news to myself but I plumped for sharing the burden three ways. "Do you remember the wall-fall theory, Old Greybeard's idea that The Pit is getting bigger because great plaques of rock, thousands of tons in weight, keep splitting off the wall and falling to the bottom?"

"Yes," said Jimmy suspiciously.

"Well, when I went up the route this morning, I traversed out onto the face to the right of the Plaza Bolivar. I got to a corner and found it was a sort of huge plinth split from the main wall."

"Just what are you getting at?"

"I mean our route goes up the middle of one of those great plaques."

"Jesus. What are we going to do?"

"For a start, you can pray that the geological moment for the next fall isn't tomorrow."

9

We'd had two throws of the dice by the morning of the sixth day in The Pit. Now we would make our third, and it would be our last because we hadn't another idea in our heads about how to get out of there. Moreover, on our first two throws we hadn't really lost, we simply hadn't won. The kindly fates had intervened before we took the plunge with the helicopter or the winch. But this time, if the surface team got the ladders down in the right place and we committed ourselves to them and lost, we'd lose life and limb.

It was a cheerful thought to wake up with, I told myself, sitting up on my sleeping bag which had slid down another yard during the night towards the gaping shaft. I looked up and the great walls seemed to have closed in a little above us. The open circle of sky gazed down like a huge fixed eye, alert not to miss our violent end. The air, the rock and the jungle were motionless with expectancy. If they'd had lungs they'd be holding their breath for the big show. I wondered where Cuyak-iare and his old man were perched to get a grandstand view, and I awaited a great, booming laugh at any moment. "So, you dogs. You dared to come down here? Well, here you will remain. You cursed fools, you will never leave alive!" I rather liked that line so I put back my head and shrieked it out.

"You fools, you will never leave here alive!"

Charles and Jimmy jerked awake. "Wha-what's that?"

"So you heard it too?" I said, aghast, handing Jimmy the machete. "You go talk to him. It's the monster."

They observed me sourly for a moment then looked about them, reminding themselves where they were and what day it was. Finally Charles swung his legs out of his hammock and perked up.

"My idea is this—we get part of the winch team moving on the trail towards the white rope. Then we eat what we can, pack up, get everything up to the surface on the winch, and stand by. When and if they find the rope, they will tell us by radio and we'll start up the wall. By the time we reach Plaza Bolivar Federico should have been able to follow the rope down the 40-foot cliff and along the rim to where it's tied to the red rope. We can then maybe guide him when he lowers the ladders."

While we devoured sardines and marmalade, we worked out a neat scheme. We'd pull the fixed ropes up after us as we climbed. When we got to the plaza, we'd attach one to the red rope up there and Federico could haul it up, untie it and drop the end of the first red rope down to us again. He would then attach the rope we'd sent up to the bottom of the ladder and we could haul the ladders down to us instead of him trying to unroll them over the edge to fall as they may. Of course, there wasn't a hope in hell that it would work out that way, but it was reassuring to talk about it.

Charles reached for the telephone and I said that as it was our last day alive I'd go for the water and wash the pots. That way I could claim to be a proper goody-two-shoes when I met our Maker. Jimmy went off for what he called his Last Leak.

My first suspicion that something was amiss came when I got back from the water point to find Charles still talking on the telephone. I heard words like *radio, second camera team,* even *sound track.* Sound track?

"What's going on?" I asked Jimmy, who was briskly clearing up the bivouac.

"Cecil B. DeMille has plans for you. Big production. Three-name stars, sound, music. . . ."

Charles replaced the telephone and I advanced on him. But he put up his hand. "I know what you're going to say but you can save it. All you've done on this expedition is fix a few ropes down the wall and flail around with a machete. But I've got funds to worry about. Expedition debts. That film has to pay them. It's got to be made."

He was quite right, of course. And anyway, films should be made for the thousands of people who will never get near a Sarisariñama. So okay, I said, and what does it entail in terms of gear and damn-fool loads to carry up the wall.

"Only the 16mm movie camera and—er, a 25-pound radio to record the sound," Charles said. Sound? What sound?

"Well, the normal conversation between climbers. The technical jargon. What moves we have to make and so on. It will be great for the viewer. When the camera team zooms in on us with the telephoto lenses, our voices will come over . . ."

"Like, 'Don't strangle your bloody self with that goddamn jumar?' " I suggested.

You have to hand it to Charles. Here we were, just a few hours away from a life-and-death struggle with a 2,000-ton cornice and he was meticulously working to cover expedition expenses. There would be cameramen at the winch site, in the helicopter and even at the top of the red rope if one could be got there by Federico and Lionel. Charles would handle the 16mm camera on the way up, Jimmy would come last and haul up and coil the ropes, and Mugsy here would go first with the hammer and the ironmongery, and a fragile 25-pound radio receiver-transmitter strapped on my back. And that wasn't all.

We packed up everything, relayed it down to the winch load point and watched in ruminative silence as each load lifted slowly up that awful wall. How long it seemed since we'd done our damnedest to make that airy ride ourselves. And how utterly lunatic a project it now appeared. Before we put the last load on, a sack came down with the radio and another, smaller bag inside.

"What's that?" I shot out pugnaciously.

"Oh, that's spare film for the movie camera," said Charles. The lying toad.

We saw the last load up, hearing the murmur of the winch only intermittently through the astonishingly loud screeches of a pair of macaws who appeared from nowhere and dived and rolled around in The Pit like dolphins in a huge bowl, their vivid blue and red plumage so discordant and yet so right with the jungle setting. The telephone had gone, and now the cable went up with a stone on the end to steady it. The wall was bare again. Back at the bivouac site we cleaned up every vestige of our presence until there was nothing to show that men had ever been there. We stood fitfully around in that strange, uncluttered place counting our alternatives. Either we'd be back before day's end to clutter it again for God knows how long, or we'd be up and gone and away the hell out of this hole. Or we'd be at the bottom for good under a great tomb of rubble and rock.

The walkie-talkie crackled and we crouched round it. Lionel's voice came through excitedly. "We've got it! We found the white rope. Federico is following it down to the red rope. You'd better start up the wall."

"Okay. We're on our way."

Charles opened his pack to put in the walkie-talkie, some film and then . . . the Object. So *THAT* is what had come down in the last winch load.

"Charles," I said, "what the blazes is it?"

"That, you bloody *inglés,* is a smoke bomb. A big mother smoke bomb. It goes off with a bang, a fizz of sparks and great clouds of thick, oily, heavy red smoke."

Yes, I knew all about it. I had tried to light a fire with one once, after three days in the sopping jungle without food.

"And what do you intend to do with it?"

"I'm going to set it off at some point up the wall where we are visible to Camp Two and the helicopter so they can zero in with the cameras." Cecil B. DeMille was at it again.

"That means on the wall above the Plaza or, by God, on the bloody final rope itself?"

There is no stopping Charles when he's determined to go ahead.

"That is right," he said emphatically.

I remembered lying on the wet jungle floor next to that hissing bomb, three years before, eyes streaming, gasping my lungs out. "Well, I hope it chokes you," I said.

With Cuyakiare listening, I should never have voiced that unamiable yen.

We stood up and checked each other's harness. Then, with a last look at the bivouac, we turned and dropped down into the undergrowth. We made that familiar half-circle round the water point and immediately picked up the trail I had cut the day before. Tensed up for the big struggle above, we moved with more than usual care and quiet. First, we wanted to conserve everything we had for the final pitch, and second, we quite consciously wanted to avoid ruffling you-know-who's feelings. So we proceeded circumspectly, like someone visiting a church of another religion, not knowing what's holy or not, or what cannot be stared at or touched. Soon we came to the bottom of the climb and I put on my best lost, glassy look, staring around vacantly.

"Wh-which way now, fellows?" I said.

"*Up,* you bloody *inglés, up* and *out.*"

We climbed the first chimney together and without a rope. We went up at the same speed, without conversation, like the practiced team we were becoming. At the top I pointed to the left up the new gully I'd followed the day before. They came on without a murmur of argument and we sailed up it to reach the vertical slab of rock at its top. We traversed to the right along a ledge, up a little wall festooned with wet, green moss, on up a gully until I waved upwards to a dark vertical cleft and the red rope which hung down it.

"There's the first hard pitch. You must make only a direct downward pull on the rope, otherwise, if Grey-beard is right, you'll pull an 800-foot plaque off the wall and we'll be on it, falling backwards to crash . . ."

Jimmy seized me by the shoulder. "Did you ever see the weight of three men pull off a great slab of rock?"

"Sure. You can pull up on a flake and the whole thing hinges outwards, breaks off and falls with the most tremendous. . . ."

"Never mind the horrors. I want to know. Could our weight and movement start this whole thing heeling over?"

I was about to keep the clowning going and jump up and down a bit, as if to shift the wall on purpose. But despite myself, I thought better of it. And I thought too of the final 200 overhanging feet, leaning backwards into The Pit, and the split in the rock that I had seen behind it.

"I don't know, Jimmy. I imagine the whole thing would fall in its own good time. It might be today and it might be in a century."

We thought about it a little. Then Charles tapped the transmitter on my back. "Well, if we feel it falling we'll switch this on. It'll be the biggest sound effect since the H-bomb."

We reached the chimney and climbed it using one jumar as a handhold on the rope. At the top Jimmy hauled the rope up, coiled it and put it in his pack. We scurried one by one up Rotten Roger, going separately so that if anyone slipped and fell he would not knock the others down too.

And there we were at Brewery Wall. Thirty uncompromising feet with thick vegetation barring its top. About ten feet from this fence was the knot joining the ropes. I pointed to the left to where I had found a way to climb up to within about five feet of this knot. It was only slightly off the vertical, but had thick clumps of moss clinging to it and the odd root draped across it. It had to be climbed quickly and cleanly, with one's weight dispersed as far as possible on all holds. "When you've done that, you clip on the jumars. When you get to the knot, stand up in the loop of the top jumar, unclip the other from the rope and clip it back on above the knot. Stand up in that loop and then bring the other jumar up. Don't hesitate to take one jumar off the rope. The other will hold you safely."

I switched off my bronzed-Alpine-guide voice and with the sneaky advantage of having done it the day before, climbed Brewery Wall and the hedge at the top with deceptive ease. About ten minutes after calling down to Charles to come up I heard a gasp, a crack of breaking roots, the slither of boots and fingernails down 15 feet of rock and the thump of a falling man landing. But there was no need to inquire after his health for Jimmy's voice rose unfeelingly from the depths.

"*Coño,* Charles. You've wiped the whole thing clean. There's not a single hold left."

Now this sort of niggling was more like us, I thought. We'd come all this way with scarcely a whisper. Happily, I bellowed out a time check, calculated to drive a monk to booze.

"All right, you two bleeders. It's eleven in the morning

and the worst is yet to come. Get your fingers out. You've wrecked the route so I'll move the rope directly over the wall and you can bloody well jumar all the way. The frigging about is over. Move up there smartly, will yer?" And so on, and so on. The most obnoxious creature in The Pit since Cuyakiare's Dad. I moved the rope over and soon it went taut as Charles's weight came on the jumars. Up he came, steady and silent, bent on bashing my head in with the transmitter, I supposed. He pushed up violently over the vegetation and collapsed on the ledge.

"Will you get the transmitter out," he gasped. I looked at him in astonishment. I have to *hand* him the murder weapon? But he was fiddling with his own pack, dragging out that knobbly movie camera. "I'd like to film this bit. I'll try to get Jimmy as he comes up. Will you talk into the radio giving instructions and guidance and so on. Make it sound natural."

Charles called the film team on the radio to couple their receiver to the recorder or whatever. He wound his camera, leaned over the edge and pointed to the microphone. "You're on the air. None of that blue language, you mother," he said. I'd never have believed it. Two of us leaning out over a 600-foot drop, Charles making a movie and me intoning into a microphone like a dance master.

"Now shift your weight onto your left foot. Good. Unclip your right jumar but don't take your foot out of the loop. Now move it above the knot. Very nice."

Jimmy, seething at the indignity, put on his cool Conqueror-of-Everest face, and came up like a veteran. As he flung himself over the barrier at the top I shouted, "Is that the best you can do, you muckle-headed twit?" Charles stared in horror until I showed him the switched-

off mike. The clowns were onstage again. But not for
long. Just a few yards to the right was the plaza, and
about 150 feet straight above it the cornice awaited us.

And so did the Final Mystery of The Pit.

But right now we had an engineering problem: how to
get the ladders down. We got ready to bellow out our
plan to the team on top, slowly, phrase by phrase. But
from the first word we were in trouble. The word, of
course, was *"Federico!"* and the echo rippled and ran
round the walls, "ico-ico-ico-ico-ico-ico." We looked at
each other in consternation. We'd never thought of
acoustics. We needed them like a hole in the water
bottles. Another try. *"Freddy!"* And round it went.
"Eddy-eddy-eddy-eddy." Then someone yelled from
above and a double, overlapping echo chattered and
flapped round the hole like a dialect from the Khmer
Republic. Confusion and disbelief. And the walkie-talkie
was in Camp One. But this was not the moment, I
decided, to drive the spike of my piton hammer into
Charlie's jugular. We tried again and again, and The Pit
threw chirps, warbles, yappings and blips at us from
every direction. It taught us quickly that some form of
shorthand was essential. "Pull-ull-ull-ull-ull rope-ope-
ope-ope-ope up-up-up-up." Pup pup pup. It would never
work. Our complicated plan would take a week to
dictate. We were, for the most eerily preposterous reason,
in a real bloody fix.

Suddenly Jimmy braced up and took over. He gestured to us to stand still and shut up. Then he leaned out
over the edge, looked up and began his message. With a
fine sense of relieved responsibility, Charles and I listened
in admiration. It was a good scheme and eventually,

looking up and dodging a shower of stones and debris, we saw a duffel bag, with a rock in it, slide over the cornice, dragging the end of a ladder with it. The bag was attached by snaplink to the red rope, and when we pulled this taut, bag and ladder slowly descended in full control. There was a bit of a holdup as Federico coupled on the second ladder, and another as we maneuvered the bag past the ledge where I'd smoked that cool cigarette on the way down. And then at last it was there, its bottom rung just reaching our ledge. The crucial moment had arrived, but we greeted it without ceremony. Charles touched the ladder and glanced upwards.

"Well, *inglés*," he said. "It's all yours."

Now, it's one thing to climb one of these wire ladders when it's up against a wall of rock, and quite another when it's hanging free. In the first case, you put your hands and feet on the rungs in the normal way. In the second, you must put your arms and legs around the ladder, seize the rungs from the back with your palms towards you, and put your feet in heels first. What Charles and Jimmy didn't know and won't know until they read this, is that I had never been up a free-swinging ladder. This was a hell of a place to learn, but it was my job to go first on this expedition and I wasn't about to chicken out now. Getting the technique so absorbed me for the first 15 feet that I thought of nothing else. Nothing, that is, except that this was a murderous form of exercise and that I could not go one rung further without falling off. I quickly clipped a snaplink from my harness onto a rung and hung there, with my aching arms dangling at my sides. Fifteen feet and I was busted. And there was more than ten times that height to go. I felt a little flicker of panic, unhitched the snaplink and struggled up another 15 feet and clipped on again, swinging exhausted in my harness. I was past the ledge now and glanced up to see the ladder soaring above me, its silvery glint diminishing with distance until it vanished against the blackness of the cornice.

My God, the *Cornice!* I stared up at it in a blue funk, holding my breath. It was like a great ogre's beetling eyebrow jutting over me, full of menace. Face to face with it at last, I realized I wasn't prepared for the situation. My brain began its computer whirr. What do I do if . . . what if it breaks off . . . how many seconds from the noise of its splitting until it hits me. . . . Hanging there, I felt the loops of my harness biting under my arms and crotch. A hundred tons of rock and rubble would wipe me off the ladder in a flash. If it came when I was clipped on, the harness would quarter me, or at least rip off an arm or two. I'd be an unidentifiable blob when I hit the bottom anyway, but I might as well be in one piece, I thought. The computer read off its conclusion: If you're clipped on when the cracking roar comes, don't look up. Unclip and await the express elevator to the basement.

The grisly decision made, I pressed on up the ladder, looking only at the rung in front of my nose. Each surge upward became shorter, and each collapse in my harness longer. The only noise in The Pit was my thudding pulse. Quite suddenly it grew darker and I knew I was under the cornice. A few feet more and I stopped dead, gaping.

Right in front of my face was smooth, pink, glistening rock. Virgin rock. A vertical strip of it six feet high, one foot wide. I felt the scrape of the jagged root-ends on each side of me and I began to intone the liturgy that the three of us are still chanting to this day. How the hell . . . who the hell . . . what the hell . . .

Above me was Federico and, reverting to normal ladder technique I forced myself up, hitched myself to a strong root and collapsed beside him, breathless, voiceless, witless. I heard him yell to Charles to start up and I hung there gesturing feebly downwards, gasping and wheezing, incredulous and dumb, unable to mouth the question.

I was out. And I'd seen The Pit's final Mystery.

But now came the Bomb Incident. I saw the red safety rope snap taut and heard Lionel grunt as he took the strain. There was a well-remembered acrid smell, and a wild shriek from Jimmy down the wall.

"He's off the ladder! He's fallen off-off-off-off-off." I jerked upright like a robot, shot down ten feet of the ladder and saw, way below in the still air of The Pit, an obscene globule of thick, greasy smoke and the hanging body of Charles swinging in and out of it on the end of the rope, puking and retching. Cecil B. DeMille had clipped the bomb to the ladder and detonated it. But without the ghost of a breeze, the smoke stayed where it was enveloping him in its evil, red cloud. Without thinking, I let go the ladder with one hand, gathered up eight or ten feet of it and then dropped it.

The weight of its nearly 200-foot length snapped it straight and the bomb broke away, still spewing smoke, to begin its long fall to the bottom. Charles was feeling blindly around in the now-thinning cloud for the ladder and after a few moments got himself weakly onto it again. Way above him I got off it quickly and swarmed up the safety rope past Federico and 20 feet further to Lionel. With Charles moving up safely again it was time to sink back into contemplation of the miracle. I was still in a trance when Charles appeared below, his face a marvelous mix of disbelief, bewilderment and delight. He came on up and collapsed beside me, his lips moving, silently mouthing the liturgy. What the . . . who the . . . how the . . . ? In remarkably quick time, Jimmy, who knew something about ladders, came up to join us from his solitary state as the last man in the hole. He stared at us and then back down at the cornice in wonderment, but we grabbed him and pulled him up the final few feet to the trail to Camp Two. It was a rough route but we tore along it nonstop, and now at intervals we were beginning to get the words out. "How?" And after another 50 yards, "Who?" And after another five minutes, "What?"

We stumbled finally into the clearing at Camp Two and soon the helicopter clattered in, gathered us up and made off south. We gave ourselves a communal bear hug, laughing, shaking our heads, throwing up our hands in hopeless surmise. We jumped out of the chopper at Camp One and stood there oblivious to the mob, shaking each other by the harness. And at last we found our voices.

Who cut the slot out of the cornice—the cleft that let us climb straight through it? Like a slice out of a cake. Like a wedge out of a cheese. Leaving that strip of clean, pink rock exposed. It wasn't Lionel or Federico or the surface team. How could they without a saw twelve feet long? And where did the great wedge go? There was no debris on the plaza. No sign of its passage down the wall. How does a slice of rock and rubble, eight feet by six feet by three feet at its wide end just disappear? Where did it go? By what physical laws was it removed? And why did this malignant Pit relent at the very end when we were wide open for the kill? What the hell . . . who the hell . . . how the hell . . . There's only one answer.

It had to be *Cuyakiare*.

APPENDIX I

Unanswered Questions

by Dr. Charles Brewer-Carías,
leader of the expedition

THE MAKIRITARE CALL THE THREE PLATEAUS WE explored: Sarisariñama-Jidi, Jaua-Jidi and Guanacoco-Jidi. The names have a certain significance in the mythology of this advanced indigenous group. The term *jidi* means simply *mountain* and corresponds to the word *tepuí* used in the Gran Sabana region and also meaning *mountain,* but in the Pemón Indian language. Jaua-Jidi is the proper name of the plateau and has no translation. Sarisariñama is the name of an evil spirit (orosha), according to one version, who lived on this plateau and used to eat men. As he ate he made a noise with his mouth that sounded like *sari.* There is also a hawk called Sari-sari, but it has no connection with the name of the mountain. Guanacoco is also the name of an orosha who lived on this plateau beneath the waterfall which tumbles down the west wall. He too was a mythological enemy of the Makiritare. There is, however, a seed with the same name, used for necklaces because of its bright red color. It has nothing to do with the name of the plateau.

Sitting on the bank of an unnamed river 5,850 feet up on the Jaua plateau, surrounded by scenery and vegetation never before seen by man, I watched the black and red waters collecting from the forests and valleys of the surface. Bubbling over the pink, almost vitreous rock of its bed the water ran north to form the Marajano River, still on the top of the plateau. It falls down the northern

171

escarpment to meet up with the black waters of the Cácaro, and then to flow into the green waters of the Erebato. This mixture of waters flows in turns into the Caura, with its yellowish waters, and eventually, carrying its particles of plateaus and jungle, into the Orinoco and thence to the sea.

Some 3,500 million years ago the water did not flow from this plateau. In fact, the process was the reverse. Waters from distant areas moved into the zone to deposit sediments from 7,500 to 9,000 feet deep at the bottom of lakes and marshes. The rains which fell continuously during millions of years when the earth's crust cooled, partly demolished this initial rock formation, breaking it down into sand. This sand settled in calm waters and, as layer followed layer, ripple formations hardened in the lifeless silence of the deep waters. At the bottom of the smaller of the two holes on Sarisariñama we found fallen rocks marked with these ripples. The lakes which formed plaques of rock are almost all in Venezuelan territory with minor extensions towards the frontiers of Colombia, Brazil and the Guianas. Because of movement in the underlying crust at some unknown time, water began to filter slowly into the sandstone aided by the enormous hydrostatic pressure of the columns of water forming vertically within this great thickness of rock, and probably by some chemical effect which does not seem to exist in today's atmosphere.

These enormous drainage torrents were channeled along lines of weakness to form subterranean rivers. Here and there, two currents of water meeting head-on, or meeting with some obstacle, produced whirlpools and vortexes which carved out domes or circular vaults. Because of erosion or meteorization of the surface of the plateaus, or because of the extension of these domes upwards to meet the surface, the roof eventually fell inward, creating the great hole we explored on Sarisariñama. In parts the walls of the holes still retain their dome shape so that the floor of the hole is wider than the opening at the top. This explains the difficulties we had to overcome during our six days of attempts to escape from the main hole.

The bottom of the hole is formed of great blocks of rock fallen from the roof and the walls. In the smaller hole there is a waterfall which spouts from one of the cracks in the wall but which has not even begun to wear away the rocks on which it falls before disappearing among the boulders to some deeper level through which ran the fossil river formed by forces that do not seem to exist any longer.

The form and sequence of the deposit of sand and other sediments seems to follow a fairly similar pattern in all the plateaus to the south of the Orinoco. This sequence of sediments has been named the "Roraima Formation" since Roraima, which sits astride the frontiers of Venezuela, Brazil and Guiana, was chosen as the typical locality for study. The same sequence is found in the three plateaus of the Upper Caura and their ages have been calculated between 1,700,000,000 and 2,000,000,000 years according to the potassium/argon and rubidium/strontium methods, and is determined indirectly by estimating the loss of these elements in the diabases which penetrated the layers of sandstone after they had been horizontally sedimented. Determining the age of the new rock penetrations gives the minimum age of the surrounding rock.

The history of vegetable and animal life, before the appearance of man, is imprinted in rock in the form of fossils in many parts of the world. These fossils have enabled us to trace the form and sequence of their development and how the species of fauna and flora which now populate the earth were derived from each other. Fossil remains have been sought for a long time in the sediments of the Roraima Formation in the hope of correlating the age of the formation with the era in which the fossil creatures or plants lived. But the search has always been unsuccessful. It was only with the use of the most modern atomic K/Ar and Rb/Sr methods that we have understood why all searches have so far failed. If the method is correct it is probable that neither life nor oxygen had yet appeared in any quantity in the earth's atmosphere when the sand was deposited.

Most fossilized plants and animals disappeared thousands or millions of years before man appeared, and some of them disappeared precisely because man did appear. Because of genetic stability or extreme adaptability to different habitats, or because they existed in places where the process of selection did not succeed in finding better substitutes, some plants and animals have continued to live into our era, and their strange forms, behavior and reproduction methods are witness to their remote past. Based on botanical collections in the massifs of the Guiana Shield, by such explorers as Cardona, Steyermark, Maguire, Dunsterville, Wurdack, Bunting and myself, it has been found that up on the plateaus the conditions of altitude, solar radiation, humidity, pH and available soil nutrients have seemingly changed very little since the appearance of life on earth and this has permitted the conservation of vegetable species now extinct in other regions. These perpetuated species are called *relicts*.

In a similar way some original species mutated and were selected by the limited environmental conditions found on each of the summits of the sandstone massifs and by the microclimates prevalent even on a single plateau. This phenomenon of evolution by selection of mutations was such that today more than 90 percent of plants found on these plateaus are found there and nowhere else on earth. Each of these plateaus is an island-on-earth where in spite of geographical proximity, environmental distances, where topography and air currents are factors, are impassable. This circumscribed regional evolution is called endemism, and is especially marked in the vegetable kingdom.

In the river which ran nearby the camp on the Jaua-jidi we found various forms of life seen for the first time on these plateaus and perhaps never before seen in other parts of the world or even classified by the taxonomists. The few forms of life in the water seem to have adapted themselves to, or remained in, conditions of pH (acidity of 4.5 of pH), abundance of electrolytes, of temperature and concentration of oxygen so that these factors have helped them to thrive, whereas other forms of life could not put up with them. As it was our objective to "seek out the origins," we were surprised by the presence, uncovered by the helicopter mechanic, of planarians, one of the most primitive animals in the whole zoological scale. These turbellarian worms were among the first animals to develop a specialized nervous system in the form of primitive multicellular eyes even though they maintained such primitive evolutionary characteristics as that of having no anus, of moving by means of cilia, of bisexual or hermaphrodite reproduction or asexual reproduction by spontaneous severing of the body.

While we were collecting a group of planarians which had hidden themselves in a crack to escape the midday sun, Bernardo, one of our Makiritare helpers, called me to watch the old phenomenon of reproduction by mutilation. There, before our eyes, a planarian separated itself into three parts, each of which made off in a different direction through the water. This demonstration of immortality left all of us in a thoughtful mood. This form of reproduction which was probably common millions of years ago has been employed by the planarian for all of what appear to be the 500,000,000 years of its existence on earth.

Also at the bottom of our river were the red algae (Rhodophyta) which appeared 2,000,000,000 years ago and were probably one of the principal producers of oxygen in an epoch when there was none of this gas in the earthly atmosphere. Nearby, a red water tick crawled among the algae's filaments. The only vertebrate we found in the river was a Gymnotiform fish. This fish is elongated like an eel and is related to the electric eel. It produces electric discharges or uses the electricity to find its way, avoid obstacles and localize its prey. It may possibly prove to be a species new to science. There was also an aquatic coleopteron, Ditiscidae, half-hidden under a ledge of rock. Among other insects which go back hundred of millions of years were the dragonflies (Zygoptera) in varied colors which fluttered about preying on smaller flying insects.

The varied habitats of the plateau's surface also produced evidence of this isolation and conservation with biological forms and methods of reproduction which were more common and predominant in other ages. The Lycopodium plant we discovered has been found imprinted in rocks of the Devonian period 370,000,000 years ago. Its complicated reproduction system goes through various stages including one where male, whip-tailed gametes swim through the water at the base of the plants seeking the female sexual organs. Among other plants we found were selaginella, which have been in existence for 270,000,000 years, hepaticae for 300,000,000 years, and lichens, a happy marriage between algae and fungi which have been on earth for 400,000,000 years. A very special plant, the Podostemonaceae, lives submerged all its life like an alga but flowers when the river dries up in its season. Finally, as an example of forms of adaptation, we were able to watch how the carnivorous plant, Drosera, trapped insects which were attracted by its brilliant red fronds.

As we have only been able to put forward hypotheses and empirical observations on origins, the need for deeper research is obvious. To help this, I would like to offer my fellow taxonomists and specialists a list of problems I have encountered, whose solution would provide some of the necessary clues to understand the evolution of life in these plateaus and the origin of the plateaus themselves:

1. Is the formation of holes and caves in the Roraima Formation a process which still continues, or are these holes signs of an activity long dead? I incline to the latter view.

2. Why are living relicts so frequent? Is this an objective observation? Is it due to the fact that frequent mutations are not produced, or to the fact that conditions have not notably changed since these forms of life appeared?

3. Why are there species common to all the sandstone formations of the Guiana Shield (general endemism of the zone)?

4. Why are there genera endemic to the Guiana plateaus, only represented by species endemic to each

plateau and even by several on each plateau (*Navia* genus)?

5.   How to explain the presence of plants with world or continental distribution or such different origins in time as lichens and orchids? Is this due to a form of successful propagation?

6.   And what of some intercontinental families such as the Rapataceae with representatives in South Africa and Australia alone? Does their endemic presence in these plateaus mean the continents were joined when the family was distributed before the fragmenting of the Gondwanaland Continent?

7.   What interrelations of flora and fauna could be statistically processed to determine the chronological sequence of the origin or the separation of these plateaus from other continents or from one another?

These questions and many more can only be answered by expeditions penetrating the remote regions south of the Orinoco. The process should be one which I like to call "constructive adventure." In other words, funds and energy should not be expended on adventure alone; it must be combined with serious field research. There is the work of a lifetime for explorers and scientists alike in that strange and beautiful zone. I, for one, intend to devote all my resources and the rest of my active life to the unraveling of its mysteries.

SPELEOLOGICAL COMMENTS ON THE EXPEDITION

by Dr. Eugenio de Bellard-Pietri,
of the Venezuelan Academy
of Physical, Mathematical and Natural Sciences

THE DISCOVERY AND EXPLORATION OF THE SARISARI-
ñama Plateau has unquestionably opened a fabulous new
chapter in world speleology. This giant plateau is, as far
as has been investigated by reputed geologists, a nearly
solid sandstone mountain belonging to the Precambrian
period. This means, amongst other things, that this
colossal formation is somewhere between 1,400,000,000
and 2,000,000,000 years old, which makes it one of the
oldest in the world.

Sandstone has not been known so far to yield large
caverns; these are found mostly in limestone and dolo-
mite, a rare exception being lava caves, also called lava
tubes. For those familiar with limestone landscapes, a
karstic region is invariably attached to a vast limestone
area where vertical shafts or potholes alternate with
dolines, poljes, vanishing rivers, very large springs, caves,
great fissures and similar phenomena typical of limestone
country. Dolines and poljes are vast depressions which
sometimes pocket the landscape, the poljes being ex-
tremely large closed valleys usually with incoming and
outgoing creeks or rivers.

Well, no matter how incredible it may seem to veteran
speleologists, this classical limestone panorama, geologically identified as a karstic region or plateau, is precisely
what met my astonished eyes on flying over the savagely
beautiful Sarisariñama-Jidi on December 1, 1973. Two

179

enormous potholes or vertical shafts, plus a number of huge dolines, crevasses, intermittent lakes, diaclases and faults, made up a formidable and strange panorama, awesome in many ways, matched as it was against a colossal and forbidding emerald green jungle which covered most of this zone with "shoulder to shoulder" density.

A number of low-flying passes made over this part of Sarisariñama in December 1973, and during the expedition of February 1974, have led me to the following speleological conclusions, reinforced after the expedition's successful descent to the bottom of the principal abyss, and the smaller one to the south.

In the first place, it is my firm belief that there exists a tremendous system of caverns under the apparently solid and massive plateau. Such a formidable network of vast passages can be visualized by noticing the formidable size of both potholes, opened by a complex solution-erosion-abrasion process; the incredible number of diaclases and smaller fissures that fracture the sandstone; the spectacular size of numerous depressions and of the dolines farther southeast; the astounding volume of water contained in both intermittent lakes, which are sometimes drained dry and at other times are completely inundated.

Such geographical phenomena, plus the knowledge we have of the uncommon rainfall that deluges Sarisari-ñama nearly all year round, point to the fact that only an outsized subterranean drainage system can dispose of such volumes of water. Such a system is called in speleological jargon a *karstic apparatus* or, more simply, a karstic drainage system. Doubtless, such a mammoth network of underground passages must have numerous inlets accessible to human beings and some equally large outlets: resurgences and cavern openings. We trust the latter will be eventually found and will permit the safe exploration of this magnificent sandstone mountain. It is my personal belief that these resurgences will be eventually found on the bluffs and perpendicular cliffs of Sarisariñama, most possibly hidden by the dense vegetation that surrounds the foot of the plateau. When explored, such a network of magnificent caverns should easily prove to be one of the largest cave systems of the American continent. These underground waters in all probability feed the mighty Caura River.

As to the mechanism that formed the two enormous potholes, a small number of theories could be forwarded in order to give a reasonable geological explanation. I nevertheless am inclined to reason as follows:

In the distant past, many hundreds of millions of years ago, water seeped down the numerous fissures in the sandstone plateau and finally arrived at a level where vertical progress was halted by impermeable nonfissured rock or clay, at a very deep level. From here on, water progressed horizontally via other fissures, called *joints*, until it finally appeared at the foot or bluffs of the mountain. Free flow of increasing volumes of water gradually enlarged these vertical and horizontal conduction tubes. Two of the largest vertical fissures possibly occupied the hearts of the present-day potholes.

This constant enlargement of the horizontal water ducts, very possibly under extreme pressure conditions when rainfall was at its peak in the region, was accelerated as time went by and the channels increased in diameter, by turbulent erosion and violent abrasion, the latter resulting from the destruction of the sandstone lining the passages and consequent scraping of the walls by resulting stones and boulders.

The undermining of certain sectors of the plateau by

the vast enlargement of the waterways directly beneath must have obviously caused larger vault collapses and, therefore, new water turbulence in the area due to obstacles stopping the free flow of water. A final collapse due to the instability of the vaults and the constant vertical seepage through crevasses opening on the plateau, are responsible for the opening of the immense abysses on Sarisariñama.

The largest of the two explored is, as far as available speleological literature in July 1974 goes, the largest natural vertical shaft so far known in the world. There are a few deeper in Mexico, France and Greece, but none so vast and wide.

Communications and Equipment

by Lionel F. Jugo

As the expedition's radio team, federico isaias and I had to set up four communication links:

    a)   Base camp to Caracas

    b)   Base camp to forward camps

    c)   Base and forward camps to the helicopter

    d)   Camp Two to the bottom of The Pit.

The original radio base was to be in Sta. María de Erebato but we changed this later to Camp One on Sarisariñama. This move gained us altitude but cut down our space for rigging aerials. After our final move to Camp Two at the edge of The Pit, space was even further reduced to what we could cut out of the jungle. Getting there, by the way, with 15 cubic feet of packing crates containing 160 pounds of delicate equipment, brought problems of its own.

The equipment for each link was as follows:

    a) Single sideband Stoner, SSB 150, 125 watts, adequate for daytime and early hours of evening before foreign stations to the south, particularly military, started up and crowded the air. It has four crystals allowing for frequencies 8470, 7490, 3570 and 6730. This last was our 24-hour standby with Fundasocial in Caracas. This set also allowed contact with the helicopter at Cacurí, for example, outside the range of the normal ground to air equipment.

    b) Single sideband Stoner SSB 21, 20 watts,

single channel, using a crystal for frequency 6730. In emergency this could contact Caracas in daylight hours. It was in Camp One in charge of Archie and our doctor, Chucho. We switched on hourly in case there were messages and made contact anyway each day at noon.

     c)   COMCO transceiver, VHF, aeronautical (123.6mh).

     d)   Military field-telephone set TA-312/PT, and Regency Standby 1 walkie-talkie on VHF aeronautical frequency (123.6mh) via which the team in The Pit could also connect with the COMCO set.

Apart from this we had a reserve set, a Stoner (military) PMC 12, 20 watts, on the same frequency as the main set. The antennas were dipole, half wave. Power for all sets came from three marine-type batteries, 70 amperes/hour. All equipment was selected to work on 12 volts. Besides these, all sets except the SSB 150 had their own batteries. We used a Honda 300 generator to charge the batteries including those shuttled over from Camp One by the helicopter. All plugs and connections were standardized to facilitate any equipment interchange. This equipment may appear excessive but we took it all because we had to guarantee communications under conditions which were unknown. During the first ten days in Sta. María and on Sari all went well. Then our troubles began. The major problem was humidity, almost 100 percent in this region. The first set to fail was the main Stoner 150 which began to emit vivid blue sparks—very attractive, but deadly for the set. I tried to repair it on the spot but the damage was too serious. We managed to combine a helicopter shuttle and a flight out of Sta. María to send it to Caracas and we got it back in one week, by which time we had returned to Camp One. In the interval we used the military PMC 12 which stood up perfectly to the humidity, giving further proof of the value of military equipment on this sort of expedition. We found too that the COMCO was sufficient for communications with the helicopter, the two camps and the walkie-talkie in The Pit.

A major problem in jungles is the lack of space for antenna. At one stage we had it strung over the heliport clearing and had to dismantle it hurriedly every time the chopper flew in. Later we found it preferable to mount it at the side of the clearing with a mast made of saplings. This lost us a little in performance, but it was worth it. One piece of equipment which proved invaluable was a simple system of plugs which permitted the use of a single antenna for four frequences.

In the main, our long-distance contact with Caracas concerned fuel. When the drums were emptying in Cacurí we would warn the Air Force to drop further supplies off by C-130, and then inform the helicopter it was on its way. We could also tell the pilots how the weather was on Sari so they could come straight there if it was clear instead of going to Sta. María.

The person-to-person calls to Caracas and from the press in London and other parts which so astonished the skeptical *inglés* were not a matter of magic but of juggling. We would simply hold the handset of the field telephone to the microphone of the radio and remember to switch off the speak button of one and switch on that of the other, and vice versa, depending on who was talking. A more tedious problem was carting the handset 200 feet to the winch site each day and then bringing it back. But as if this wasn't enough we had to run the winch as well. Long before the expedition Charles had asked me to get or design a machine which was light, and which would lift 250 pounds up a height of 1,000 feet. This required a lot of study, mainly in speleology journals, and I ended up with a 3hp motor with a specially made frame and drum. Getting it to the edge of The Pit and setting it up was a struggle. We staked it down on a foundation of tree trunks then rigged an eight-foot trunk to stick out over the hole with two pulleys on the end to keep the rope clear of the cornice. We would sit astride this trunk to look down into the hole (but only when we had to!). The three men down The Pit say they heard the motor as a faint murmur, but for us the din it made in the total silence of the plateau was nerve-wracking. Another difficulty was getting the loads off the rope once they arrived under the tree trunk. And when The Pit team put on loads which stuck out beyond the metal cone, we had problems with them catching on the overhangs.

The machine worked well enough, but the fibers of the two nylon ropes kept splitting and catching in the coils on the drum. We found the best way to clear the obstruction was to burn them off with cigarettes. We were very much against the idea of trying to bring a man up and were relieved when the tremendous crash of the falling bag of rocks dissuaded The Pit team from trying it. If there is a next time—and I hope there is—we will be better prepared for it.